TECHNIQUES FOR INDUSTRIAL POLLUTION PREVENTION

A Compendium for Hazardous and Nonhazardous Waste Minimization

TECHNIQUES FOR INDUSTRIAL POLLUTION PREVENTION

A Compendium for Hazardous and Nonhazardous Waste Minimization

By Michael R. Overcash

Professor, Department of Chemical Engineering
North Carolina State University
Raleigh, North Carolina

Translation by
Michelle L. DeHertogh
Texas Instruments
Houston, Texas

From
New Processing Techniques of the Industry of France

La Mission Technologies Propres de la
Direction de la Prévention des Pollutions
du Ministère de l'Environnement

LEWIS PUBLISHERS, INC.

121 S. MAIN STREET, P.O. DRAWER 519, CHELSEA, MI 48118

Library of Congress Cataloging-in-Publication Data

Overcash, Michael R.
    Techniques for industrial pollution prevention.

    Includes the translation of part of Les techniques
propres dan l'industrie française.
    "Materials from New processing techniques of the
industry of France by la Mission Technologies Propres
de la Direction de la Prévention des Pollutions du
Ministère de l'Environnement."
    Bibliography: p.
    Includes index.
    1. Factory and trade waste—Handbooks, manuals, etc.
2. Hazardous wastes—Handbooks, manuals, etc. 3. Conservation of natural resources—Handbooks, manuals, etc.
I. Techniques propres dans l'industrie française.

English. Selections. 1986. II. Title.
TD897.5.084   1986   363.7'28   86-2722
ISBN 0-87371-071-1

LEWIS PUBLISHERS, INC.
121 South Main Street, P.O. Drawer 519, Chelsea, Michigan 48118

PRINTED IN THE UNITED STATES OF AMERICA

Dedicated
to
Mary, Rachael Tabitha Yoong,
Mother and Father,
and
David

This document was originally assembled and developed as:

Les Techniques Propres dans l'Industrie Française
by
La Mission Technologies Propres de la
Direction de la Prévention des Pollutions
du Ministère de l'Environnement

The original document in French has more complete technical descriptions of each process summarized in this translated English version. Any reader wishing to order the complete text in French should communicate with:

Ms. Florence Petillot
Chef de la Mission
"Technologies Propres"
Ministère de l'Environnement
14 Boulevard du Général Leclerc
92524 Neuilly sur Seine Cedex
France

PREFACE

This book is a compendium of successful waste elimination schemes. The technologies described, along with the chapter outlining the overall approach to handling industrial wastes, are applicable to a variety of disciplines. Industrial personnel will find the process changes described herein helpful in the review of possible waste reduction alternatives.

For compliance with federal and state regulations (specifically, RCRA) emphasizing hazardous waste reduction, use of this book by an individual plant will allow a statement that a review has been conducted. Compliance with the spirit of the regulations to require consideration of waste reduction is achieved if a formal review is conducted by a knowledgeable staff, regardless of whether the outcome is adoption of a waste reduction scheme or a conclusion that no reduction is feasible.

This compendium allows quick access to a large number of modifications, and is thus a sound basis for such formal reviews. From the perspective of a regulatory staff, access to a large number of possible approaches can assist in suggesting waste reduction techniques. These benefits to industrial and regulatory personnel also apply when nonhazardous effluents and emissions are reviewed for minimization options. Though waste reduction is not always economically feasible, the proven processes described in this book will be very helpful in many cases.

Michael R. Overcash
North Carolina State University

ix

INTRODUCTION

A program at North Carolina State University (School of Engineering) has been underway since 1972 in research and development to eliminate or reduce industrial waste effluents and emissions. This area is broadly referred to as waste reduction, waste elimination, source control, or pollution prevention "pays" (PPP). PPP requires substantial qualification, but nonetheless is a widely used acronym encompassing this field. The waste reduction program at North Carolina State University is quite parallel in concept and in certain of the technical aspects to the major governmental effort in France by La Mission Technologies Propres de la Direction de la Prévention des Pollutions du Ministère de l'Environnement. This French approach is translated and printed in the latter sections of this book.

Differences, however, exist between our objectives as a university involved with the development and implementation of waste reduction technologies and the broader goals of a government body attempting to stimulate this field and protect the environment. The interest of North Carolina State University has been aimed at establishing the actual technical basis for PPP and the linkages which might span all types of manufacturing as regards waste minimization. Nevertheless, there exist such substantial similarities in the field of pollution prevention that when we became aware of the extensive French government documentation of these industrial techniques, a copy of the French report was obtained. This was entitled "Les Techniques Propres dans l'Industrie Française," or "Clean Technologies of French Industry." In reading the report and studying the data contained therein, it was felt that there would be broad United States interest in such materials. For this reason, we undertook the translation of portions of the French report. These comprise the bulk of this book, and include excellent preface and introductory sections from the French report which were also translated. The translated materials from the French document contain valuable data with which to improve the detailed understanding of industrial waste reduction as a technology.

In translating the document, the authors have also undertaken substantial editing. The report by the French Ministry of the Environment contains 370 pages, but we decided to translate only selected elements of the typical waste reduction technical entry. The French description of each process contained 1) history and background, 2) the clean process (description with some operating variables), 3) balance sheet or evaluation (pollution, materials, energy, and costs), and 4) conclusions. We have summarized these individual process descriptions to include 1) the industrial situation in which the process modifications were made, 2) the actual flowchart of old and new processes (highlighting the specific changes), and 3) tables providing comparative economics and waste reduction values. Since the French report did not provide complete details relevant to actual design or engineering specifications, it was felt that the brief description given in this translated edition was sufficient. In addition, the Ministry of the Environment report contains sections on twenty-four processes in the development stages. These were not translated. However, if the reader wishes to explore the further written material on a specific process or the 24 research stage processes, the entire report in French can be procured from the French Ministry of the Environment (address given in the front of the book).

CONTENTS

WASTE ELIMINATION TECHNOLOGY
BY
PROFESSOR MICHAEL R. OVERCASH

Since 1972 the interest in waste elimination as a technology for managing industrial wastes has steadily developed. The written material in this chapter is predominantly the work of the author in response to a need of the general public, government personnel, and industry for information on the broad scope as well as specific details regarding waste elimination technology. These materials have been used in courses, technical presentations to corporate staff and professional societies, and as an introduction or aid with regard to procedures for implementing waste minimization plans.

Emphasis on source reduction of wastes has received a major governmental focus in the field regulated as hazardous wastes. Initiatives in many states have also focused on hazardous wastes. The use of waste reduction is however primarily justified in a comparison to the cost of the alternatives for treatment (such as incineration) or ultimate disposal. Thus the economic magnitude of hazardous waste management serves as a major influence on the growth of PPP activities. However, industry at present spends substantially more on the treatment and discharge of water-based effluents (6-fold) and atmospheric emissions (7-fold) than on hazardous wastes (U.S.Department of Commerce 1985). In light of these qualitative expenditure differences it is clear that waste minimization should be stressed at least equally for air and stream discharges as for hazardous wastes. In fact, a substantial number of waste reduction approaches for hazardous chemicals and wastes are derived from research and development related to air and water pollution control. In this context, waste reduction for the hazardous waste field is neither unique nor of a quantum level greater in complexity than the general industrial efforts to reduce waste emissions.

RELATIONSHIP OF WASTE MINIMIZATION TO INTEGRATED HAZARDOUS WASTE MANAGEMENT

The detailed discussion which follows is focused on hazardous waste because of the large national emphasis toward the search for alternatives to landfills. For nonhazardous industrial wastes the same alternatives exist; and with air emissions there is a parallel set of choices regarding treatment and allowable stack discharges. In the wastewater field the choices are between elimination and treatment versus stream discharge. Thus the reader should interpret the role of pollution prevention in the broadest context of the entire range of industrial wastes and discharges.

Schematically, hazardous waste management is appropriately viewed as a hierarchy of alternatives, Figure 1. Each level in this hierarchy has been carefully designed to emphasize the major philosophical, policy-making, and technical decisions that are involved in the selection of alternatives for any given industry. This hierarchy schematic and the designations of perpetual storage, conversion technologies, and in-plant options were first developed in the form of Figure 1 as a part of a detailed environmental course and book (Overcash and Miller, 1981).

There are basically three general options: minimization or reuse of the hazardous waste, conversion of the hazardous waste into nonhazardous or less hazardous material, and perpetual storage. These general options differ substantially in terms of philosophy, time-frame, technique, and economics. At present, perpetual storage methods are the most prevalent for hazardous wastes and, hence, are the focus of attention for both regulatory and industrial personnel.

Optimum, successive utilization of the three general options would result in a decreasing amount of hazardous wastes as treatment proceeds in the direction of the arrows in Figure 1. Only small volumes of the more inert materials should be considered for a perpetual storage. The hierarchy identified in Figure 1 represents a preferred path in the management of hazardous wastes as follows: (1) in-plant options should be used to reduce the volume and toxicity of generated hazardous wastes, (2) wastes that are generated should be converted to less hazardous forms and should be reduced in volume, and (3) the remaining residues and wastes that are hazardous should be stored in a manner that minimizes risks to the environment and the public.

Identification and understanding of this hierarchy permits rational decisions to be made, lower-risk solutions to be identified, and better public understanding of the issues. The management approaches outlined in Figure 1 also stress that no single approach is sufficient to solve all hazardous industrial waste management problems. Further, the concepts identified emphasize prevention and reduction rather than treatment and storage; and emphasize the goal of reducing the risk to the environment and public. Thus, the emphasis is on positive rather than negative management options.

Waste Minimization Options

In-plant options are probably the most effective and economical means of managing hazardous wastes. These options represent approaches that generally minimize impact on public health and the environment. Reducing or eliminating waste production is substituted for end-of-pipe management. These in-plant options include:

1. Process modifications to eliminate or reduce the volume of

2

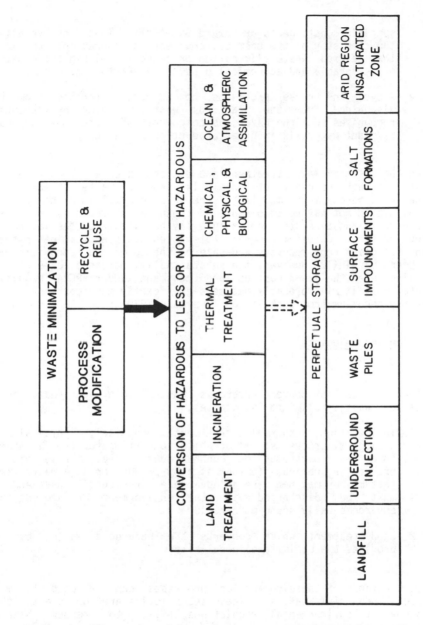

Figure 1.   Hierarchy of hazardous waste management alternatives
(Overcash and Miller, 1981).

specific constituents of hazardous wastes. These involve altering the chemistry or the chemical engineering operations to achieve the desired waste elimination or reduction within the constraint of acceptable and economical product manufacturing.

2. Recycle and reuse processes to prevent materials from being discharged from the plant as waste. Use of these techniques recognizes that the hazardous components of wastes may be usable reactant materials in other production processes.

As the costs of waste treatment and disposal increase, cost-reduction achieved through recovery of chemicals from a waste stream prior to disposal will become more important. The value of reclaimed chemicals plus the savings in not having to dispose of these wastes could match or exceed the original disposal costs. An example may be cited from the leather and tanning industry. Chromium removal from wastes has been well demonstrated, but the net gains from recovery and sale on the commercial chromium market have been marginal. However, the substantially increased costs of secure landfill or even the lower cost of land treatment under RCRA regulations indicate that it will probably become significantly more cost-effective to remove the chromium from such wastes (Overcash 1980).

Conversion Technologies

Technologies that convert hazardous wastes into less hazardous or nonhazardous wastes fall into two classes:

1. Incineration, thermal treatment, chemical, physical, and biological processes, all of which convert wastes from a hazardous to a less hazardous or nonhazardous state. These processes produce a residue (either as a by-product or as a waste stream) that may or may not have an adverse environmental impact and that must be discharged to the environment or stored in an environmentally sound manner. B

2. Land treatment which converts the hazardous wastes but also provides the ultimate disposal site.

Combinations of these conversion processes can be used to manage hazardous industrial wastes. These technologies are consistent with the philosophy of environmental regulations (i.e., to render hazardous industrial wastes nonhazardous).

4

Perpetual Storage

Perpetual storage is the most prevalent existing hazardous waste management practice. Each perpetual storage technology attempts to place the waste material in a highly condensed or concentrated configuration in which the hazardous constituents do not move. Generally little or no conversion from a hazardous state occurs and hence care, monitoring, and migration prevention are required for an indefinite period. A National Academy of Science committee decided that at least 500 years was realistic as a period of concern for hazardous wastes in landfills and perpetual storage options (Natonal Research Council 1983). Regulations under RCRA establishing the period of concern for landfills as 30 years should be considered to be unrealistic. Careful attention to design for containment over short post-closure periods does not eliminate the probability of containment system failure after 30 years. Instead, the likelihood of adverse impacts and the realistic costs of perpetual care by compliance to the letter of the law (closure under RCRA) are only masked.

The storage technologies clearly involve a long-term obligation because with time, particularly at a closed site, most of the changes that can occur are adverse (e.g., eventual penetration of a surface cover, gas diffusion and leakage to the atmosphere, and leakage of mobile constituents to groundwater). As an illustration, experience with some nuclear waste repositories provides an example of unsuccessful results with long-term storage. Despite the best intentions these have a proven record of leakage to the environment during a relatively short period amounting to a small fraction of a 500-year lifetime which may be a realistic period for perpetual storage.

The central questions presently being addressed with regard to perpetual storage options concern the technology and the procedures for implementation. The question still not being widely addressed is how to accommodate these sites over a probable greater-than-500-year life. It is indeed possible that hazardous waste problems and consequences are merely being postponed and will have to be dealt with by future generations. If this could be argued convincingly (or even addressed), then a realistic consideration of the direction for hazardous waste management could be undertaken.

Time Frames

The options noted in Figure 1 have different liabilities and time frames associated with them. A very approximate depiction of the substantially different time frames over which various hazardous waste options can be expected to have an exposure impact is given in Figure 2. The in-plant options relate to very short time frames (hours to days), with

5

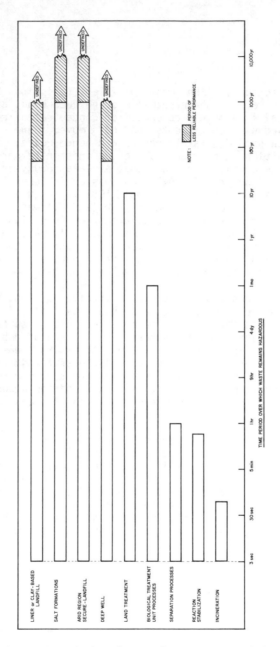

Figure 2. Approximate time frames for waste remaining potentially hazardous within the alternatives for managing hazardous waste.

effects limited to those of short duration (i.e., those resulting from extreme operating conditions). They are subject to normal community acceptance and Occupational Safety and Health Administration considerations.

The environmental and societal effects of conversion technologies in Figure 2 involve an expanded time frame (minutes to years). Short- and long-term effects on air and groundwater quality must be addressed. Dispersion occurs in a current time frame and is taken into account as part of the design and monitoring approach. These effects must not violate acceptable air, drinking water, and land use standards. Thus, the effects of these technologies are essentially short term.

For the storage category in Figure 2, the time frame is much longer (decades to centuries). The liability is therefore likely to be shifted from those who generate or store the waste (the present generation) to future generations. These technologies could provide a potentially adverse environmental legacy.

For technologies in which the character of the hazardous waste remains for a long period (30 years) there is increasing likelihood of noncontainment. Reliable containment for very long periods (1,000 years) is a disadvantage when the true costs of monitoring and remedial action are included with the placement of a waste in such facilities. The shorter the lifetime of the hazardous waste technology, the better the control and the lower the dispersion potential generally associated with that technology. Longer storage periods create much larger unknowns and much less opportunity for management and control.

The discussion of waste reduction in relation to the overall management of hazardous or nonhazardous waste would not be complete without the statement of an overall objective for environmental protection. If one examines the perpetual storage segment of Figure 1 or the concept of stream discharge or atmospheric emissions it is clear that these all represent the environment. Discharges to the environment cannot be forced to zero. A nonzero discharge concept is firmly based on both thermodynamic and economic principles. In addition the environment is not without a capacity to treat, detoxify, or otherwise assimilate waste emissions of all types. This concept must be recognized and a balance achieved between the various societal factors which are present. Utilization of the nonzero capacity of the environment to assimilate discharges to soil, groundwater, surface water, ocean, and the atmosphere should be adopted and would represent societal nonsense not to do so. It should also be recognized that the environment capacity will vary among the various discharge media and is surely substantially smaller than the total of waste generated by our society.

Thus in Figure 1 one must recognize that the material flow to the environment (perpetual storage for hazardous waste residues) is nonzero but potentially a small percentage of the absolute mass of waste generation. There are two broad alternatives for achieving the reduction necessary to match the acceptable material flow to the environment. These are treatment technologies and waste minimization alternatives, Figure 1, and are relevant for wastewater, air, or solid waste emissions. The goals of environmental efforts should thus be to use these two reasonably distinct approaches to select the most cost-effective means of achieving desired discharge to the environment. The goals should not be to force one approach, such as waste eliminationchieve cost-the necessary reduction between total waste produced and final discharge.

In the context of the above arguments there are several consequences which must be recognized with respect to pollution prevention,

1. Waste elimination cannot be a requirement to the exclusion of alternatives for managing hazardous or nonhazardous wastes. This is important since public and governmental pressure will exist to simply mandate waste elimination.

2. In many cases it is less expensive to achieve the acceptable levels for environmental discharge through treatment technologies rather than by means of waste reduction techniques. In this regard this chapter is in disagreement with the philosophy of the French compendia contained in this book (see translated Foreword section by Environment Ministry of France).

3. It must be recognized that there will always be residues unless massive economic committments are made to reuse all materials. To a certain extent the inevitability of residues should focus attention on the management of such materials, i.e. perpetual storage in Figure 1. In particular the challenge is to determine the true cost for long term storage of hazardous wastes. This cost will be necessary to adequately establish industrial fees and in fact to permit industry to make an economically effective decision among waste elimination, treatment, or perpetual storage. With the true cost of perpetual storage established there is a legitimate case for deciding to put any type of waste into such alternatives. This is a subtle argument and area of consideration since it leads in a direction counter to the major flow of environmental regulations, and yet is a recognition that the entire field of waste management is interrelated.

WASTE MINIMIZATION TECHNOLOGY

Waste elimination, as well as other areas of hazardous waste management, is subject to ambiguity as to exact definition and whether a particular case is a waste elimination scheme. For example, incineration is frequently viewed as a technique for the treatment or detoxification of hazardous wastes. However, when certain side-streams in industry have a

8

high solvent content, then instead of drumming and placing in a landfill these streams can be burned for energy content. In this latter case, the incineration would be viewed as waste elimination via recycle/reuse. As another example, metal recovery (electro-winnowing) can be used as :

1. a technique for selectively purifying a process stream, thus serving as waste minimization;

2. a technique for treating electroplating wastewaters to meet industrial discharge standards to POTWs or to detoxify a hazardous waste, thus providing end-of-the-pipe treatment;

3. a technique to recover selected metal sludges placed in landfill cells, thus being an integral part of the perpetual storage phase of hazardous waste management.

These ambiguities are reduced by keeping clear that in most cases a) it is easy to designate the specific usage of a particular technology in this hierarchy, Figure 1, and that b) the hierarchical system is actually a decision-making process as well as the simple classification of individual technologies.

Subcategories Of Waste Elimination Technology

Understanding of the relationship of waste minimization approaches to the broad hierarchy for hazardous waste management, it is appropriate to next explore the technologies that comprise pollution reduction . Two reasonable separate subcategories are included in the following description of waste elimination.

- in-plant waste modifications

- recycle/reuse techniques

A technology separation is necessary here since a number of the characteristics of these approaches are different, since there are often quite different advocates of each, and since it is likely that the barriers and solutions will differ partially between these subcategories. The following (along with Table 1) are a partial description of subcategory differences.

The in-plant modification category is characterized by direct linkage to the main income-producing operation of the plant. Thus, changes (intentional or unintentional) have a potentially substantial impact on a major corporate cash flow. This can be an advantage or a liability, depending on the effect of changes to eliminate wastes. The emphasis in

TABLE 1

COMPARISON OF IN-PLANT MODIFICATION AND RECYCLE/REUSE APPROACHES TO WASTE REDUCTION

IN-PLANT	RECYCLE/REUSE
DIRECT LINKAGE TO MAIN INCOME PRODUCING OPERATION	OFTEN STREAMS CONTAIN DIVERSE CONSTITUENTS
EMPHASIS ON PREVENTION OF OCCURRENCE	EMPHASIS ON SEPARATION OR REGENERATION
TYPICALLY, STREAMS ARE LESS DIVERSE IN CHEMICAL COMPOSITION	CAN BE REMOTE FROM PLANT; HENCE FURTHER ECONOMIES OF SCALE
INVOLVE MODIFICATION OF EXISTING EQUIPMENT (LESS EXPENSIVE)	NOT LINKED STRONGLY TO FLOWS AND CONDITIONS OF MANUFACTURING
CONSTRAINTS CAN BE SUBSTANTIAL BECAUSE OF PRODUCT QUALITY OR HISTORICAL APPROACH	

this category is on _prevention_ of waste occurrence. As a rule the waste streams are less diverse at this early manufacturing stage so that the potential for predictable changes is improved. Capital costs for in-plant modification are generally lower than for recycle/reuse since the major process investment is already present.

The recycle/reuse approach is more often used on streams that have been combined or modified and hence have more diverse constituents than found directly in-plant. An emphasis is placed on separation or regeneration to yield a usable material. Recycle/reuse can be located remote from the plant, hence economies of scale are possible with more than a single source of waste entering the facility. This separate character of these processes essentially unlinks recycle/reuse processes from the rapid variations in flows and concentrations associated with direct manufacturing processes. The reader should keep in mind the above differences between in-plant process modifications and recycle/recovery when exploring the broader field of waste minimization. The approach to the assessment, selection, and adoption of alternatives which lead to waste reduction is reasonably similar for both of these subcategories. In the following section this generalized approach is described in detail.

IMPLEMENTATION OF WASTE MINIMIZATION TECHNOLOGIES

There has been an initial tendency by many to characterize the waste elimination field as extremely diverse, highly proprietary, and lacking in a unified technical approach. Thus, a perception develops that little is transferable among industries and that reinvention of basic principles is required in most cases. However, on closer review of a number of actual case studies and utilizing engineering process theory it is clear that there are certain substantial elements of waste elimination that are transferable and therefore serve as the technical base for waste elimination concepts. These common elements were developed (Overcash and Miller, 1981) with the objective of establishing the recurring facets of waste elimination so that more rapid utilization of these experiences can occur across industry and geographic lines. Therefore, it is important to note that _waste elimination is much more a thought process or problem-solving sequence which attempts to go further back into the source of waste and to employ engineering principles which reduce hazardous waste or recover useful materials from such wastes_.

One of the objectives of the waste reduction program at North Carolina State University has been to establish a unifying theory or organizational structure for the entire technological field of waste reduction. This overview approach was felt to be necessary to accelerate the broad adoption of pollution prevention through an improved understanding of these technologies. This basic approach has been coupled with research and development of specific instances of waste reduction in industry. Specific waste reduction schemes are also being documented on a nationwide basis at

11

an accelerated pace as emphasis in this field is increased.

The first question with regard to the entire field and technology in 1980 was how to organize the disparate information that existed. An organizational structure to encompass the industrial approaches to waste reduction was first studied along three alternative patterns. The first was centered on the pollution prevention practices within a particular industry. Published literature has a major tendency to cluster along these narrow industry-specific lines. As a consequence innovation in an area such as solvent recovery in the chemical industry can be slow in recognition or adoption by another industrial category, such as textiles. Within industrial groups the issue of proprietary concern is not uncommon, and this was felt to potentially limit the full potential for a broad adoption of waste reduction techniques.

The second pattern was to view all forms of industry as containing a set of generic operations or sources of wastes. In this way, the similarities across industrial lines are emphasized and waste elimination approaches can be interpreted in a more generic fashion.

The third pattern is based on the chemical engineering concept of the unit operations as the elements which actually comprise all manufacturing operations. Examples of these unit operations are continuous or batch reactors, extraction, heat transfer, distillation, and mixing. At this level of organization the description of waste elimination would involve details of design criteria, flow rates, operating conditions, and the multitude of chemical engineering consideratons that are a part of industrial operations. This third organizational pattern is the most detailed and correspondingly has substantial proprietary restrictions with regard to information access.

For the purpose of education and the accelerated transfer of technology, the generic organizational scheme was adopted. With this approach the issues of which areas of manufacturing had been most profitably addressed for waste reduction or had yielded the greatest mass or toxicity reduction could be studied across many industrial categories. Proprietary considerations were substantially reduced so that access to information was enlarged. In addition other advantages discussed below were achieved. There were also limitations with this middle-ground approach. A generic evaluation leads the engineer or designer not to specific process variables, but only a suggested approach. Thus the industrial personnel must undertake detailed implementation on a case-by-case basis.

The generic concept for investigation of pollution prevention is based on the observation that actual process modifications or recycle/reuse techniques cluster around four areas of manufacturing. These four areas, Figure 3, are:

- chemicals recycled for process use

12

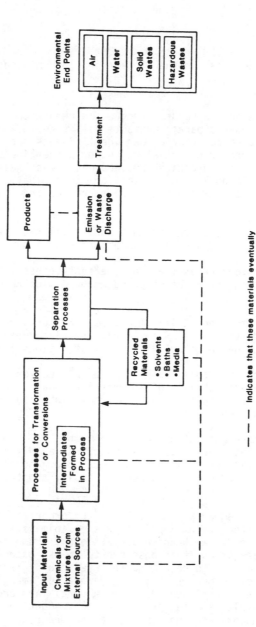

Figure 3.　Generic manufacturing sequence indicating areas subject to waste reduction or recycle/reuse technologies.

- intermediates, whether intentional or unintentional

- product

That is, when one reviews a hundred such waste reduction case studies, the solutions tend to have separate characteristics depending on which of the four generic areas of manufacturing was identified as the problem origin. This clustered character of actual pollution prevention solutions was the reason that the generic organization approach was felt to accurately reflect the overall pollution prevention technology. The purposes for selecting this generic organizational framework for the study of pollution prevention were:

1. transferability, since for most industrial facilities these four areas are common, and hence solutions can be more widely used

2. with these four areas the cost effectiveness and magnitude of waste reduction can be compared among a wide number of industrial groups

3. assistance to industrial managers and engineers can then be structured to focus on the more effective areas of manufacturing in order to begin actual design investigations for waste reduction

4. as experience in this broad field of pollution prevention is gained the availability of an organizational structure serves as a basis for refinement to improve transferability and acceptance of waste reduction

5. finally since automated design of chemical and other product manufacturing facilities is maturing rapidly it is important to have pollution prevention in a compatible general format so that waste reduction objectives be more easily adopted by industry

In summary, a framework with which to organize and improve the implementation of the technology of waste minimization was selected as illustrated in Figure 3. With an organizational framework which covers most industrial categories it is possible to investigate a large number of specific case studies and to determine the areas of greatest success in pollution prevention. Conversely, the areas of marginal or negative economic gain can be identified and in general avoided by industrial personnel responsible for achieving waste reduction. That is, the organizational framework and subsequent analyses represent the theory of pollution prevention as a broad technology. In order to undertake such analyses a large number of case studies would be needed. It was for this reason that the authors undertook the translation of the Environment Ministry of France compendium for industrial waste elimination procedures. The translated compendium in reduced form is contained in the later sections of this book.

WASTE REDUCTION PROCEDURES

A detailed review of a substantial number of waste elimination case studies indicates that rather than being an extremely diverse collection of technologies there is a fairly unified and predictable pattern. It is this unified pattern which is the transferable basic principle which should be understood in order to efficiently undertake and encourage the implementation of waste elimination. This pattern or set of principles is actually the collection of phases or stages through which the evaluation or decision process moves, rather than the more diverse results of such a decision process. If cost-effective, this sequence will usually lead to implementation of an in-plant process modification or recycle/reuse scheme.

For both in-plant process modification and recycle/reuse approaches, there are four stages common to the usual implementation of such a waste elimination scheme. These are as follows:

1. identification of the chemical(s) of concern in the waste or emission;

2. establishment of the origin of the chemical(s) of concern;

3. selection of the technically feasible reduction or recovery techniques;

4. economic comparison among waste elimination alternatives and among other approaches in the hierarchy for waste management.

In waste elimination the above stages represent a procedural approach that is reasonably transferable among industries and thus is a general format for such technology.

The first stage is identification of the chemical constituent(s) within the waste that is of concern, whether hazardous, priority pollutant, air or water quality limit, etc. This identification process is meant to sharpen the focus on the specific portion of the waste stream that creates the primary environmental impact, whether from a regulatory or an actual effects perspective. A focus on specific waste species is important since more precise selection of elimination or recovery techniques is then possible. As a contrast, fewer alternatives exist if the task is instead wholescale reduction of the volume or mass of a hazardous waste. Such reduction usually still leads to a hazardous waste, since volume reduction by water removal is the prevalent technique. As a partial observation, the elimination of specific constituents is often more cost-effective since the process techniques can be more directed than is the case for alternatives to reduce total waste.

15

The second stage is the determination of origin for the chemical constituents identified in stage one. From past experience the four groups which emerge as useful in establishing origin are illustrated by dashed lines in Figure 3. The four groups are:

1. reactant or input material;

2. internal reuse or recycled chemicals;

3. intermediate compounds (intentional or unintentional);

4. final product.

These four groups have been selected and refined from extensive evaluations of actual control or recovery case studies. That is, the elimination techniques cluster distinctly into these four groups, thus allowing the industrial personnel a more direct procedural selection process from among all available techniques. The emphasis is on the transferable approach since process selection is more generic when viewed as being related to the presence of these four waste constituent groups common in most industries.

A third stage in the process of waste elimination is the selection or evaluation of technical feasibility for methods that reduce the emission of the specific constituent(s) of concern. Solutions can best be described in generic terms for each of the four categories of origin developed in stage two and illustrated in Figure 3.

For reactants or input materials, consideration should be given to different starting compounds and chemistries for reaching the final product. In some cases, a contaminant in the current input materials is the problem, and reformulation by the supplier can eliminate the compound(s) of concern. In addition, conservation or greater conversion to final product leads necessarily to less starting material that can appear in the waste and thus reduces waste generation.

In the case of internal reuse chemicals such as solvents or carriers, the substitution of chemicals with improved physical properties (lower volatility, lower water solubility, etc.) or with less toxic character are primary methods of waste elimination. A second technique is to improve the separation methods so that more of the reused chemical is actually recycled.

Intermediate compounds can typically be reduced by altering the reaction conditions or residence time presently used in a manufacturing process. To a certain extent, this source of waste constituents is directly tied to the primary production process and often has only limited options for elimination in-plant.

The fourth group, the final product from manufacturing, is managed primarily by improving separation processes. That is, if product is appearing in the waste stream, then better separation into saleable form is usually the available approach.

The final stage in the process of waste elimination is economic comparison among:

1. the technical alternatives developed to directly reduce or eliminate the specific chemicals of concern in an industrial waste, and

2. the alternatives to waste elimination, as illustrated in Figure 1, such as treatment/conversion techniques or perpetual storage options (with due consideration of environmental equivalence).

Cost-effectiveness is thus the final factor in implementation of a waste elimination project. This economic facet is emphasized continually in the published literature of waste elimination and signifies the attractiveness of a detailed evaluation in almost all industrial hazardous waste situations.

An inherent advantage in the preceding approach is that it provides a transferable data base and procedural methodology for the field of waste elimination. Experience in elimination or recovery of a recycled chemical can thus be transferred between industries. This transfer is most effective in the initial technology assessments since it is obvious that the exact use of a waste elimination scheme must always be refined in any particular plant. These modifications are chosen on both technical and economic-effectiveness bases. Such refinements are inherent in any manufacturing process .

UNITED STATES WASTE REDUCTION STATUS

On the basis of published literature, industrial discussions, the assessments of national committees (described below), and ongoing conferences in the pollution prevention field it is reasonable to assess the current waste minimization status in the U.S. This assessment also weighs the amount of Federal and State funding for this field in comparison to that in waste treatment or landfill areas. The present status (1982-86) is that a major effort in waste minimization, across diverse categories of industry has not been undertaken. If one depicts the present and future trends in waste elimination technology, Figure 4, an S-shaped curve results. This shape is typical of a number of phenomena which undergo substantial growth but ultimately reach a stage in which further waste elimination is not incrementally cost-effective. In the early phase a minimum cost must be incurred to achieve any waste reduction. This cost involves an improved understanding of the need for waste reduction often through increased management understanding. Preliminary evaluations are needed to determine the presence of problems. Then relatively small investments lead to very substantial levels of waste reduction, the

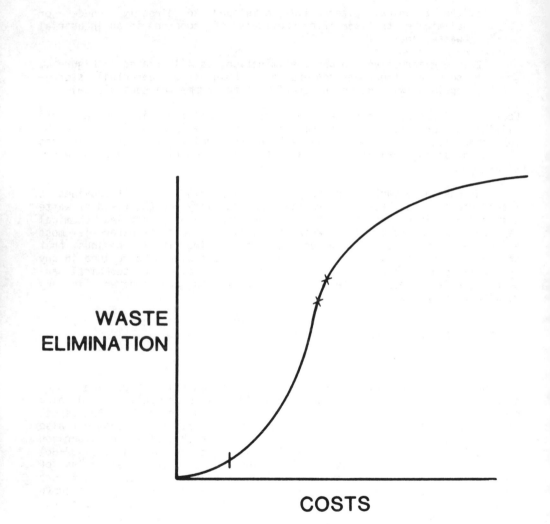

Figure 4. National industry relationship between waste reduction or
recycle/reuse results and the costs incurred to implement
these alternatives.

intermediate segment of Figure 4. Finally the objectives of very high levels of recycle or waste elimination lead to greatly increased implementation costs. In this latter segment the economics of treatment rather than process modification or recycle/reuse may be more attractive.

For a period of 10 months a committee of the Environmental Studies Board in the National Research Council undertook a detailed examination of reducing hazardous waste generation (National Academy of Science 1985). The conclusions from these deliberations represent an additional assessment of the current United States position and future trends for waste reduction. Their conclusions were as follows:

1. Most waste reduction efforts in U.S. industry are still in their early stages. Many opportunities exist for reducing the generation of hazardous waste. Efforts should begin now to encourage industries to take advantage of these opportunities.

2. At the current stage of development of industrial waste management programs across the nation, substantial progress in reducing the amount of hazardous waste generated can be achieved by employing relatively simple methods that entail modest capital expense. Such methods emphasize engineering or plant specific circumstances. The amount of waste generation that can be avoided is, unfortunately, not known, because of difficulties in obtaining reliable data.

3. The current trend toward increasing costs of land disposal for hazardous wastes--through greater liability for generators and site operators as well as through restrictions on the use of land--is an extremely important impetus to implementing waste reduction programs. To encourage reduction in the amount of waste generated in the future, this trend bringing the cost of land disposal to the level of its true costs to society should continue.

4. An important impediment to implementing low-cost waste reduction practices is lack of access to information about them. Developing means to exchange and disseminate information about successful waste reduction projects is an essential first step toward reducing future waste generation.

5. Approaches other than the direct regulation of manufacturing processes are needed. Within the regulatory framework, regulations would be beneficial that are administered consistently and predictably and are flexible enough to encourage the use of methods that reduce the generation of hazardous waste.

6. In the long term, as implementation of newer, more capital-intensive technology becomes necessary to reduce waste generation further, public policies will need to adapt to the

19

different considerations. Industry may require assistance--in the form of incentives or subsidies, for example--to help defray R&D as well as capital costs. Risk assessment and risk management will be needed to balance the trade-offs between protection of public health and the environment and costs. Because of the time required to develop and implement new technologies and conduct risk assessment programs, research on these topics should begin now.

SOURCES OF WASTE MINIMIZATION INFORMATION

Several books and reports are available to those responsible for the evaluation of technical options for waste minimization. Since industries generating hazardous waste will be required to document that waste reduction was evaluated, these technology sources will prove essential. The evaluation process should include a survey of available technologies in the context of a particular plant to decide that

1. certain waste minimization techniques are appropriate and were adopted leading to some quantifiable reduction in hazardous waste generation or

2. the evaluation of available compendia of alternatives led to no appropriate or cost-effective waste reduction

Either decision complies with the spirit of the RCRA amendments of 1984. It is important that the reader note the emphasis should not be on waste elimination to the exclusion of acceptable treatment. That is, the appropriate national objective should be the reduction of residues which ultimately must be stored in the environment. That objective is achievable through waste elimination, treatment technologies, or combinations of these. The combination should be dictated by economics, given that reasonable environmental protection is achieved, and not by an arbitrary requirement to adopt hazardous waste reduction.

Compendia of waste reduction alternatives are an efficient means of information searching. Such sources for waste reduction are listed below. In addition journals on hazardous waste, chemical engineering, and specific industry or trade business represent the major publications for the waste reduction alternatives.

1. Royston,M.G., Pollution Prevention Pays, Pergamon Press, New York 197p., 1979.

2. Huisingh, D. and V. Bailey,ed., Making Pollution Prevention Pay, Pergamon Press, New York, 156p., 1982.

20

3. Campbell, M.E. and W.M. Glenn, Profit from Pollution Prevention, Pollution Probe Foundation, Toronto, Canada, 404p., 1982.

4. Huisingh, D., H. Hilger, S. Thesen, and L. Martin, Profits of Pollution Pevention, North Carolina Board of Science and Technology, Raleigh, N.C., 197p., 1985.

5. Noll, K.E., C.N. Haas, C. Schmidt, and P. Kodukula, Recovery, Recycle, and Reuse of Industrial Wastes, Lewis Publishers, Chelsea, MI. 196p.,1985.

6. Tavlarides, L.L., Process Modifications for Industrial Pollution Source Reduction, Lewis Publishers, Chelsea, MI., 150p., 1985.

7. Huisingh, D., L. Martin, H. Hilger, N. Seldman, Proven Profits from Pollution Prevention, Institute for Local Self-Reliance, Washington, D.C., July 1975.

LITERATURE CITED

1. U. S. Department of Commerce, Pollution abatement costs and expenditures, 1983, Current Industrial Reports, MA-200(83)-1, U.S. Govt. Printing office, Washington, D. C., 83 p., Apr. 1985.

2. Overcash, M. R. and D. Miller, Integrated Hazardous Waste Management, Today Series Text, Amer. Inst. of Chem. Engr., N.Y., 580 p., 1981.

3. Overcash, M. R., The Evaluation of Waste Management Options for Leather and Tannery Hazardous Wastes, Biological and Agricultural Engineering Department, N.C. State University, Report to U.S. Dept. of Agriculture Eastern Regional Research Center, 231 p., 1980.

4. National Research Council of National Academy of Sciences, Management of Hazardous Industrial Wastes: Research and Development Needs, Publ. NMAB-398, National Academy Press, Washington, D.C., 77 p., 1983.

5. National Academy of Science, Reducing Hazardous Waste Generation, Nat'l Academy Press, Washington, D.C., 76 p., 1985.

FOREWORD

(AS TRANSLATED FROM FRENCH REPORT)

"...to produce better while polluting less..."

Such are the words which challenge industry, facing more economic and environmental protection requirements. The development of new techniques in the French industry prove that these two ideas may not be contradictory but rather convergent and complementary.

Indeed, French people expect an increased effort from industry as far as the pollution generated by the factories is concerned. Meanwhile, the high cost of the energy and the scarcity of raw materials compel industrialized countries to review their basic structures and their modes of production. As a result, a technologic change appears in order either to transform the production tools (which were old fashioned or not updated and therefore face the international competition which turns out to be tougher) or to perfect new fabrication technologies in order to look for some new and necessary channels for trade and exportation.

The action of these factors must favor the integration of pollution prevention and of environment protection at the very heart of industrial production units.

The focus of this catalog is to show clearly that "deeds, not political words" have occurred in France, as follows:

o 73 techniques or processes have been implemented in 121 factories;

o 21 techniques or processes are being developed industrially which shows the continuation of both research and innovations. (These are not a part of this translated book.)

This list is neither exhaustive nor complete. However, the industrial panorama which is given is timely showing the diversity of the industrial areas, the pollution problems that occur and the solutions which are proposed.

The credit for the results which have been obtained has to be given to the manufacturers themselves who had the will to choose the new technologies; the economical balance-sheets show they were right as they have tried to conciliate the struggle against pollution and industrial profitability. Most of the time, an effluent or emission which is not controlled corresponds to a waste of either raw materials or energy.

23

This development has also been made possible thanks to a system of incentives based on new requirements and financial aids.

I hope that this catalog will inform the factory managers about the technical capability as far as original solutions to environment problems are concerned.

Thierry Chambolle

Head of the Pollution
 Prevention Section
Environment Department

PREFATORY NOTE

(AS TRANSLATED FROM FRENCH REPORT)

Some of the new techniques introduced in this catalog are not in service yet. The list if not exhaustive and must therefore be regarded as examples. Most of the techniques which are described have already been put into service in the industry; however, this report also includes the techniques whose development has not been completed yet and which are based on recent studies and research. The 73 charts have been dedicated to the techniques which have already been put into service in the factories. The new process is compared to the old one.

For each example, the results which are shown focus on a specific factory where the new process has been established. The factory indicated is not necessarily the first to have chosen the new process, nor the one which has obtained the best results. The purpose of the catalog is not to make any comparisons among factories but to report actual situations which enable one to provide factual and not estimated data, especially in terms of pollution.

At the end of some of the examples are mentioned some factories which also chose the new process. The process examples have been classified by activity area: food industry, construction materials, chemicals, metallurgy, surface treatment, tannery, wood and paper paste.

The format and the structure of the examples are always the same. The new process is introduced along with the industrial context in which it takes place. A description goes through the details of both old and new processes and shows clearly the technical feature of both. The description goes with a picture on which the characteristics of the new process are drawn in blue (shaded in this translated book). This description is completed by some comparative balance sheets:

o in terms of the pollution of both processes (old and new);

o in terms of materials and energy needed by both processes;

o in terms of investment expenses and working costs.

The most significant elements are printed in blue (shaded in this translated text). In the economic balance sheet, unless otherwise stated, the costs do not include taxes. The investments do not take into account the cost of the building site and the operating costs include neither the constant expenses (insurance, general expenses) nor the redemption of the investment.

The duration of this redemption is sometimes mentioned when it is a technical redemption. The main conclusions result from the comparison which shows the advantages of the new process along with its possibilities of improvement and extension. The reader will find at the end of this catalog a Glossary of the parameters used to describe the pollution along with the symbols mentioned in these example processes.

25

INTRODUCTION

(AS TRANSLATED FROM FRENCH REPORT)

POLLUTION

Unfortunately, this meaningful word is used more and more in conversation and in the newspapers. As for the responsibility, the manufacturer is guilty, of course.

However, industry creates employment and products. It actively participates in the increase of the national income. But no one notices it when the problem of pollution arises.

Indeed, the nature of industry is not well known by the French people. Who knows what are the raw materials and the processes that are used to make the products of everyday life?

Too often, factories seem only to have:

o Effluents in rivers which lead to the death of fish;

o Bad/dirty smoke that contaminates houses and agriculture;

o Disgusting smells;

o Noise from machines.

MEANS TO FIGHT AGAINST POLLUTION ALREADY EXIST

Manufacturers know quite well they have to take care of the environment. If not, they will have to pay fines and taxes, and people will be against them. Two behaviors are possible. The first one is to admit that pollution is inevitable. This leads to the addition of anti-pollution devices to equipment already in place. These devices treat pollution at the output and are called external treatments. A lot of materials, processes, products have been suggested by specialized industries for a long time. They treat by chemical, physical, or biological means. They are:

o equalization stations for wastewater

o filters for atmospheric pollution

o incineration for residues

The implementation of such equipment generally is expensive and the results aren't always sufficient, but remain necessary for a lot of industries. So this first behavior is easy to set up, available at any time, but is expensive and difficult to operate.

27

MORE THAN POLLUTION CONTROL: ANTI-POLLUTION

The second behavior is very different. New techniques are established to suppress pollution at the origin, not at the output. They try to stop the creation of pollution, usually by three possible arrangements:

1. Cleaning and maintenance of workshop. How to take good care of water, energy, raw materials, products. For example, separation of fluids polluted by others, to avoid useless and expensive dilution; limitation of waste; loss of raw materials, products; bad cleaning; etc.

2. Modification of fabrication processes. These are new techniques that don't change the nature of the process in itself but reduce the pollution. For example, separation and recovery of materials that were supposed to be discharged; put the workshop in closed circuit; separation and increased value of by-products.

3. New processes. The logic of production is changed with an important benefit for the environment. One substitutes in the manufacturing process a less polluting process. The irreversibility thus achieved is the absolute guarantee of limiting pollution forever. For example, mechanical decoking of steel wire; fabrication of H_2 from H_2 peroxide instead of Na hypochlorite dyeing in solvent.

This classification in three levels shows the variety of operations one can gather under the generic name "New Technology". It shows clearly the relativity of the notion of cleanliness. Indeed, cleanliness of a technology has to take into account:

o The state of the art of techniques in any one field at a given time.

o The economy of the firm, of the product, compared to the international market.

o The effects on people.

Priorities can be set up, depending on pollutants, conditions of natural resources, level of reversibility of damage done to the national environment. Nevertheless, one has to be aware that to set up a new technology is not enough to insure its success. Once the decision of changing the process is taken, everyone in the factory has to participate in its application. Indeed this will imply some changes in habits of workers, whose aware and active participation, of course, is needed.

STRONG TECHNICAL & ECONOMIC ARGUMENTS

Avoiding the creation of pollution seems to be the best way to fight against it, technically and economically.

1. Technically. On one hand, the antipollution system being integrated with the manufacturing process absorbs automatically the short or long-term variations. So we have constant reliability. On the other hand, it is always easier to act on a pollution as soon as it occurs, before it becomes mixed with other polluting products, coming from other sources. Thus efficiency and good output are maintained.

2. Economically. New technologies are generally interesting from the economic point of view of the firm. If we forget pollution, the setting up of a new process is often profitable in itself. Moreover, if we look at the savings made with taxes and other expenses due to pollution, only a few balances are negative. Also, as pollution is to be eliminated, it is equitable to compare the cost of prevention to the cost of investment and operating that would imply equipment without identical efficiency.

PREVENTION OF WASTE

New technologies can be adapted to fight against waste, also. Taking into account the limited amount of some natural resources and the necessity of developing a recovery of materials that previously caused pollution, these technologies are good examples of economy for raw materials.

As a conclusion, one can say that each time it is feasible to do an operation by means of new technology, it is beneficial to the industry.

I. PARAMETERS DESCRIBING POLLUTION

Measures of water pollution:

o <u>BOD$_5$ (biochemical oxygen demand)</u>. This indicates the amount of O_2 consumed during 5 days by micro-organisms, especially to perform the destruction of the degradable constituents in the effluent. The method of measurement is standardized (standard AFNOR N.F. T 90 103)

o <u>COD (chemical oxygen demand)</u>. This parameter represents the amount of O_2 which must be provided by powerful chemicals in order to oxidize the pollution contained in the effluent. The COD is expressed by a figure which is larger than the BOD$_5$. The COD/BOD$_5$ ratio shows the bio-degradability of the effluent. The closer to 1 it is, the more easily bio-degradable is the effluent. The measurement method is standardized (standard AFNOR N.F. T 90 101).

o <u>The MO (oxidizable materials)</u>. It is a pollution amount which involves both COD and BOD$_5$ such that:

$$MO = \frac{COD + 2BOD_5}{3}$$

o <u>MES (suspended materials)</u>. When the effluent is filtered or centrifugated, one recovers materials: the MES. It is a non-dissolved pollution and is easier to eliminate. The measurement method is standardized (standard AFNOR N.F. T 90 105).

o <u>Equitox (Eq)</u>. This is a unit of measure for inhibitor materials. A wastewater contains as much equitox as the number of times it has been necessary to dilute it so that a "Daphnia" (small crustacea living in soft water), is no longer immobilized by the toxic and inhibitory materials contained in the effluent. The measurement is standardized (standard AFNOR N.F. T 90 301).

o <u>pH</u>. This parameter indicates whether the effluent is acid or basic. If it ranges between 1 and 7, the effluent is acid; if it is equal to 7, it is neutral; and between 7 and 15, it is basic. It is forbidden (by law) to discharge effluents whose pH is not in the range of 5.5 to 8.5. The measurement method is standardized (standard AFNOR N.F. T 90 006: colorimetric method

31

or N.F. T 90 008: electrometric method.

o Soluble Salts. The amount of soluble salt contained by wastewater can be estimated by measuring the conductivity of the water, which can be expressed in Mhos/cm. The amount of rejected salts is therefore given by the product of the volume of rejected water times the conductivity. The measurement method is standardized (standard AFNOR N.F. T 90 111). The measurement of the conductivity, according to the nature of the dissolved ions, enables one to obtain their concentration. One often writes in mMho (milli-Mho) to express an amount of dissolved salts.

o Color. The color of the waters is measured by comparison with a reference coloration scale (platinum-cobalt scale or Hazen scale). For a given volume of effluent, it is expressed by the amount of some determined chemical mixture (Cobalt chloride and potassium chloroplatinate - Pt/Co) that is necessary to be put in the same liquid volume in order to obtain the same color. The measurement is standardized (standard AFNOR, N.F. T 90 034).

Measures of Air Pollution:

o Dust. The amount of dust or solid particles in a given volume of air or gas is expressed in g/m^3, i.e. in gram per cubic meter of air at the normal conditions of temperature and pressure ($273^{\circ}K$ and 1 atm). For piped emissions, the measurement is standardized (standard AFNOR N.F. X 44 052).

Measure of Noise Pollution:

o Decibel. The level of acoustic pressure of a sound is measured in decibels (dB). To account for the frequencies that can be heard by human ears, one has defined the decibel (A) or dBA which is the level of acoustic pressure measured through the filter A of the sound meter.

II. SYMBOLS - UNITS OF MEASURE - ABBREVIATIONS

m = milli 10^{-3}	j = day
k = kilo 10^3	l = litre
M = mega 10^6	m^2 = square meter
A = Ampere	m^3 = cube meter
$^{\circ}Be$ = degree Beaume'	M = micro
$^{\circ}C$ = degree Celsius	p.p.m. = part per million
F = French Franc	PCI = Inferior calorific capacity
g = gram	s = second
h = hour	TEP - ton-Equivalent-Petroleum
hl = hectolitre	Wh = watt-hour

Kappa index: index of delignification

Abbreviations:

B. P.: Post Office Box
E: negligible
e.s : dry extract
H.T : without taxes
m.s : dry materials
MVC : monovinyl chloride
ND : not available (in tables)
NSR : non sugar reconstituted
PVC : polyvinyl chloride
  > : greater than
  < : less than

Initials:

AFB   : Financial Agency of Bassin
APC   : Nitrogen and Chemical Products
CEA   : Commissionership for Atomic Energy
CENG   : Nuclear Studies Center of Grenoble
CdF   : The French Coal Board
CTGREF : Technical Center of the National Forestry
          Commission Rural Engineering
DGRST : General Delegation of Scientific &
          Technical Research
DII   : Interstate Industry Directorate
INRA   : National Institute of Agronomic Research
IRCHA : National Institute for Applied Chemical
          Research
ITF   : Textile Institute of France
PCUK   : Chemical Products Ugine-Kuhlmann
SIDIC : International Society for Industrial
          and Commercial Development
SILF   : Industrial Society of Fala Yeast
SNCF   : National Society of Railways

NOTE ON ECONOMIC CALCULATIONS

The original French document contained cost values in francs for a variety of years in which the process was implemented. In order to convert these to U.S. dollars an average exchange rate for each year was determined from financial records of the U.S. Bureau of the Census. The values given below were thus used to express the economic evaluations in U.S. dollars.

Year	Exchange Rate (U.S. cents/French franc)
1971	18.15
1972	19.83
1973	22.54
1974	20.80
1975	23.35
1976	20.94
1977	20.34
1978	22.22
1979	23.50
1980	23.69

VEGETABLE (POTATO) PROCESSING
PROTEIN RECOVERY BY EXTRACTION

ADVANTAGES OF THE PROCESS

Vegetable wastewater treatment produces a powder that contains 80% proteins. This powder can then be used for cattle feed, in the glue and adhesive industry, and in wood panel manufacturing. Once settled, this process modification has also made it possible to reduce the organic load of the factory's

POSSIBILITIES OF EXTENSION

Improvements in the quality of the recovered product would lead to consideration for human feeding. On the other hand, trials on improving the concentration and separation are in progress with the object of increasing the value of the remaining dissolved materials and thus diminish the induced pollution from the source.

PROCESS BALANCES

Basis: ton of potatoes

Pollution balance:

	Rejects		Old Procedure	New Procedure
	throughput	m^3/t	3.5	3.5
water	BOD	kg/t	25	15
	COD	kg/t	40	30
	MO	kg/t	30	20

Economic balance: 1982 Dollars

	Old	New
Investment	2,140,980	3,347,714
Annual costs	243,055	698,783
Annual returns	0	467,710

OLD PROCEDURE

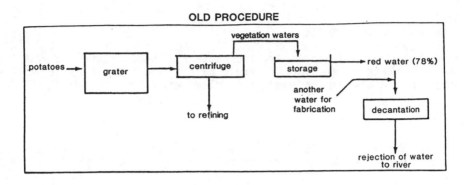

NEW PROCEDURE

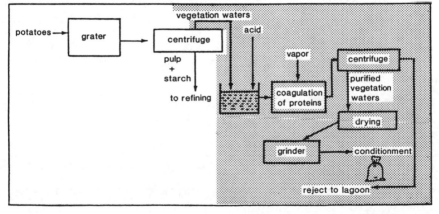

VEGETABLE (POTATO) PEELING AND WASHING
ECONOMIC RECOVERY OF STARCH

ADVANTAGES OF THE PROCESS

The recovery of starch from the wastewater used to peel and wash the potatoes makes it possible to reduce half of the suspended materials in the water and also the consumption of clean water.

POSSIBILE EXTENSIONS

The commercialization of the starch and potato wastes covers a part of the expenses due to the functioning of the biological treatment system.

Basis: ton of potatoes transformed

Pollution balance:

	Rejects		Old Procedure	New Procedure
	throughput	m^3/t	4	2
Water	MES	kg/t	35	16
	COD	kg/t	50	16

Economic balance: 1982 Dollars

	Old	New
Investment	380,015 (1)	107,933 (1)
		359,777 (2)
Annual costs	425,324	362,476
Annual returns	292,319	61,387

(1) installation of starch recovery
(2) effluent treatment station

OLD PROCEDURE

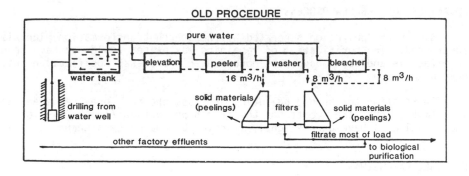

pure water

elevation — peeler — washer — bleacher

water tank

drilling from water well

16 m³/h   8 m³/h   8 m³/h

solid materials (peelings)   filters   solid materials (peelings)

filtrate most of load

other factory effluents

to biological purification

NEW PROCEDURE

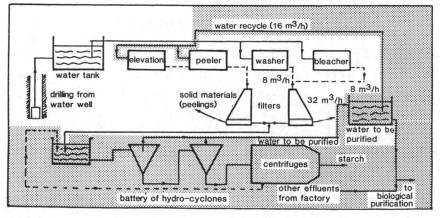

water recycle (16 m³/h)

elevation — peeler — washer — bleacher

water tank

drilling from water well

8 m³/h   8 m³/h

solid materials (peelings)   filters   32 m³/h   water to be purified

water to be purified

centrifuges   starch

battery of hydro-cyclones   other effluents from factory   to biological purification

STARCH PRODUCTION FROM VEGETABLES
COMPLETE RECYCLING OF EFFLUENT WATERS

ADVANTAGES OF THE NEW PROCESS

All of the pollution generated by the starch transformation plant is suppressed. The purification of remaining wastewater would not have been so effective and would have cost more had this process modification not been used..

In relation to the old polluting process, the new one required an investment, but the costs of the starch production (in the view of an industrial use) have been slightly improved by the process change.

POSSIBLE EXTENSIONS

The new process could be extended to similar factories; nevertheless, it is not universal. It has been found that such a process can be used in starch modification reactions (reactions that do not produce greater soluble substances in relation to starch). Thus starch oxidation into bleach cannot be carried out because it requires too much water and leads to a high solubility of the product by seriously damaging starch.

Basis: ton of transformed starch

Pollution balance:

	Rejects		Old Procedure	New Procedure
	throughput from the middle	m^3/t	6	0
Water	MES	kg/t	21	0
	BOD	kg/t	48	0
	COD	kg/t	91	0
	toxin	kE/t	0	0

Economic balance: 1982 Dollars

	Old	New
Investment	–	–
Annual costs	584,638	–

OLD PROCEDURE

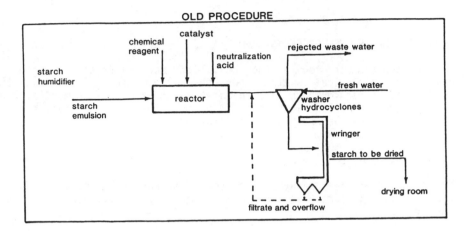

NEW PROCEDURE

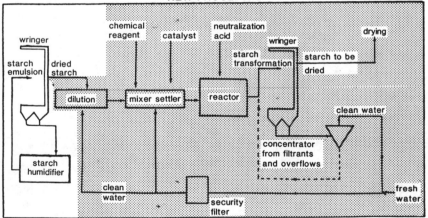

YEAST MANUFACTURING
CONCENTRATION AND UTILIZATION OF EXTRACTION EFFLUENTS

ADVANTAGES OF THE NEW PROCESS

The new process lowers the pollution more effectively than an equivalent biological system. On the other hand, it allows the possibility of recovering polluting materials that can be commercialized. This process is also less expensive, both the investment and running costs, than an equivalent biological station.

POSSIBLE EXTENSIONS

This process is already widely used. It can help many different types of yeast factories for which the effluent volume is always a problem.

Basis: ton of molasses treated

Pollution balance:

	Rejects		Old Procedure	New Procedure
	throughput	m^3/t	20	23
Water	MES	kg/t	3	0
	BOD	kg/t	130	5
	COD	kg/t	180	8.8

Economic balance: 1982 Dollars

	Old*	New
Investment	4,567,300	2,199,600
Annual costs	456,840	363,780
Annual returns		431,500

*Cost of the biological treatment station to treat the effluents from the old procedure.

OLD PROCEDURE

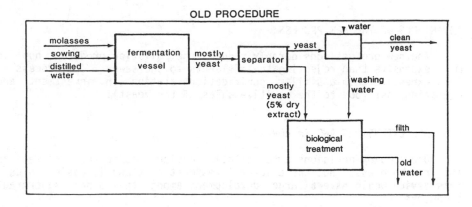

NEW PROCEDURE

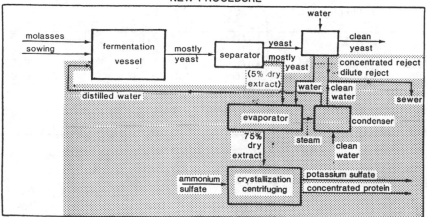

SAUERKRAUT PRODUCTION
UTILIZATION OF CABBAGE JUICE

ADVANTAGES OF THE NEW PROCESS

The new process considerably reduces the pollution produced, but is more expensive than rejecting the effluents to the sewer. The process is less expensive than a similar purification station in investment and operating cost (due to the positive sales of the yeast).

POSSIBLE EXTENSIONS AND IMPROVEMENTS

One company envisions a fermentation device that performs as well as the new process, but has a lower investment cost and is easier to use. This device would have a larger development among the other sauerkraut producers.

Basis: ton of sauerkraut

Pollution balance:

	Rejects		Old Procedure	New Procedure
	throughput	l/t	75	74
Water	BOD_5	kg/t	2.6	0.25
	COD	kg/t	4.1	1.0
	Lactic acid	kg/t	* 1.1	0.03
	Total nitrogen	kg/t	0.13	0.05
	Phosphorous	kg/t	21.8	1.5
	pH		* 3.6	7.9

*neutralizer

Economic balance: 1982 Dollars

	Old	New
Investment	7,747	9,722
Annual return	-	3,950

OLD PROCEDURE

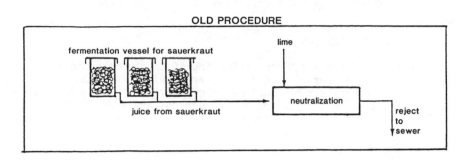

NEW PROCEDURE

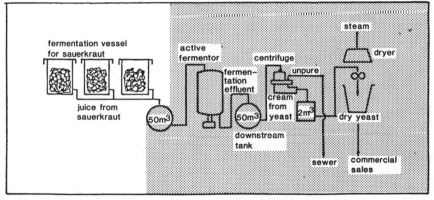

LACTOSE MANUFACTURING
RECOVERY FROM AIR EMISSIONS OF ENERGY AND LACTOSE

ADVANTAGES OF THE NEW PROCESS

Treatment of the air before releasing it to the atmosphere permits energy and lactoserum powder recovery which were previously air pollutants. The investment required is relatively reasonable and will pay for itself quickly if the energy savings and additional returns on the lactoserum are taken into account.

POSSIBLE EXTENSIONS

Such a recovery system is sufficiently profit-earning to be used by a whole lactoserum powder production plant. Many dairies have begun to use it.

Basis: ton of powder (lactose) produced

Pollution balance:

	Rejects		Old Procedure	New Procedure
Air	Temperature	$^{\circ}C$	87	45
	Particles	kg/t	10.5	3

Economic balance: 1982 Dollars

	Old	New
Investment	1,519*	68,359
Annual costs	17,539	–
Annual returns	–	9,500

*cost of reconduction of the old process

6

OLD PROCEDURE

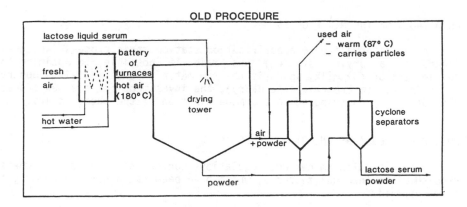

NEW PROCEDURE

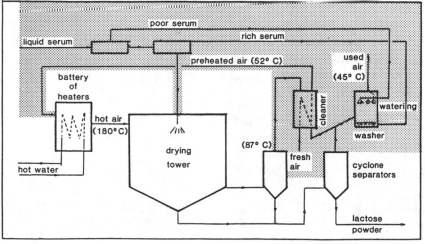

CHEESE PROCESSING
ECONOMIC RECOVERY OF RESIDUES IN CHEESE PLANT EFFLUENTS

ADVANTAGES OF THE NEW PROCESS

As an alternative to a purification station, this process (which is linked to a drying station) has made it possible to optimize the development of the milk, which is the raw material used in this industry. In this case (a French cheese dairy), the investment for the new process has noticeably contributed to the modernization of the cheese dairy.

POSSIBLE EXTENSIONS AND OTHER CASES

With attention to current regulations concerning the dairy cheese rubbish, the internal measures, which have been taken here, are going to be widespread.

Basis: ton m^3 of treated milk

Pollution balance:

	Rejects		Old Procedure	New Procedure
	throughput	m^3/m^3	2.8	2.25
Water	MES	kg/m^3	1.5	0.8
	BOD	kg/m^3	7	2.5
	COD	kg/m^3	11.5	4.5
	MO	kg/m^3	8.5	3.5

Economic balance: 1982 Dollars

	Old *	New
Investment	1,315,435	210,049
Annual costs **	175,391	-
Annual returns	-	-

*Biological treatment station capable of treating the rejects from the old procedure with 95% efficiency.

**The difference between the old procedure and new procedure from data from the sewers of the new procedure.

OLD PROCEDURE

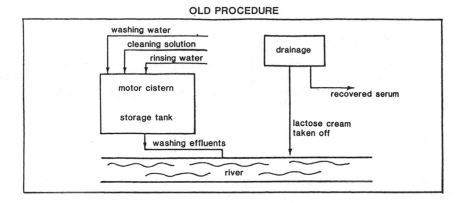

NEW PROCEDURE

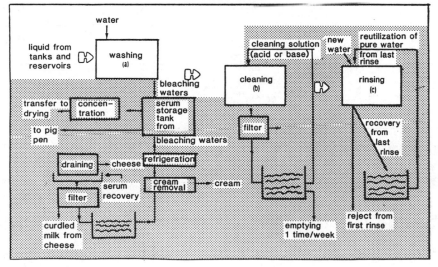

POULTRY PROCESSOR
REPLACEMENT OF WATER-BASED SYSTEM
WITH PNEUMATIC AND MECHANICAL TRANSPORT

ADVANTAGES OF THE NEW PROCESS

This mechanical and pneumatic transport process has lowered the needs for purification of effluents of the abattior by reducing the grease volume and grease concentration. For an average investment, this process noticeably reduces the operating costs.

POSSIBLE EXTENSIONS

By making small changes in the process, it can be used in many food-agronomy industries that use water as means of transport for pollution.

Basis: ton of slaughtered poultry

Pollution balance:

	Rejects		Old Procedure	New Procedure
	throughput	m^3/t	20	11
Water	BOD	kg/t	8	8
	COD	kg/t	24	17
	MES	kg/t	9	5
	grease	kg/t	2.5	1

Economic balance: 1982 Dollars

Annual costs corresponding to 13,000 tons of volatiles per year

	Old	New
Investment	20,462	58,464
	87,696*	-
Annual costs	29,232	-

* Cost corresponds to treatment plant necessary to reduce pollution output to an equivalent degree as the new process.
Remark: An investment of 120,000 F is still necessary to transport all the waste without water.

8

OLD PROCEDURE

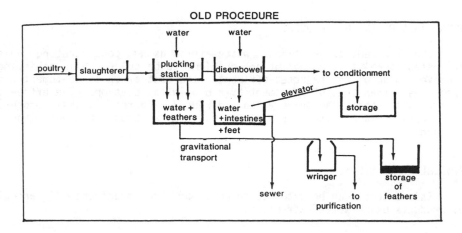

NEW PROCEDURE

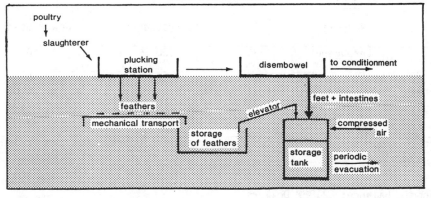

ANIMAL-BASED OIL PRODUCTION
IMPROVED RECOVERY FROM WASTE WATER

ADVANTAGES OF THE NEW PROCESS

The treatment of gelatinous wastewaters using concentrating unit processes permits the pollution problem to be solved while achieving improved value of the rejected effluent. The cost of this abatement is much less than that of treatment by a biological station. The effluent from this treatment can be recycled and reused. Therefore, this process modification makes it possible to have a workroom without liquid effluents.

POSSIBLE EXTENSIONS

This process can be extended to most of the treatment of animal by-products using water extraction.

Basis: ton of tallow

Pollution balance:

	Rejects		Old Procedure	New Procedure
	throughput	1/t	500	50
Water	MES	kg/t	2	0
	BOD	kg/t	7	0.02
	COD	kg/t	10.5	0.03
	grease materials	kg/t	2	0
Air	odors		nauseating odors	slower release

Economic balance: 1982 Dollars

	Old	New
Investment	–	232,390
Annual costs	19,000	146,160
Annual returns	–	121,900

OLD PROCEDURE

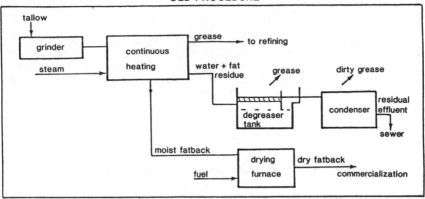

NEW PROCEDURE

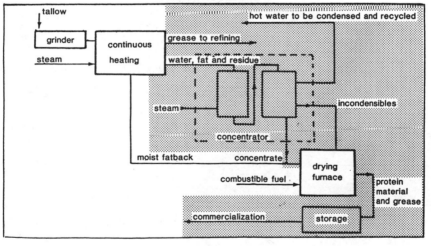

MEAT BY-PRODUCTS
GREASE RECOVERY BY CENTRIFUGATION

ADVANTAGES OF THE NEW PROCESS

From the pollution point of view, the centrifuge equipment has a definite effect since it permits pollution reduction from 60%, merely by treating the water used to scald the pork lard. The initial investment is considerably more than that of a purifying station yielding the same results. The difference can be compensated by selling the fat.

POSSIBLE EXTENSIONS

The Avon et Rayobert Company has based the setup of this centrifuge on the example of Olida at Loudeaux. So far this process is not very widespread, although it can be applied to any factory of this kind. Indeed this system is interesting since the small water content concentrated in fat is treated.

Basis: ton of paste

Pollution balance:

	Rejects		Old Procedure	New Procedure
	throughput	l/t	285	285
Water	BOD	kg/t	10	3.5
	COD	kg/t	20	8
	grease materials	kg/t	10	2

Economic balance: 1982 Dollars

	(Estimated) Old	New
Investment	43,850*	58,460
Annual costs	880*	1,710
Annual returns	-	4,180

*Estimation

OLD PROCEDURE

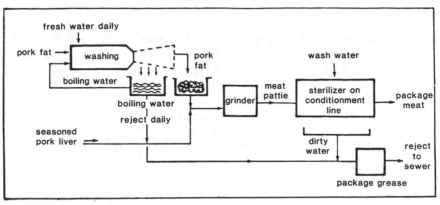

NEW PROCEDURE

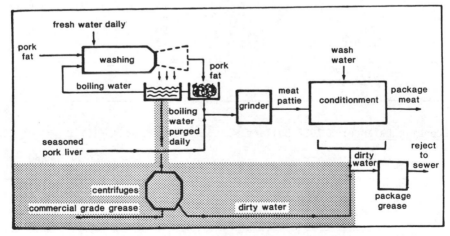

FRUIT PROCESSING
RECOVERY OF FRUIT JUICE CONCENTRATE

ADVANTAGES OF THE NEW PROCESS

It is obvious that this operation is profitable and all of the pollution is eliminated.

POSSIBLE EXTENSIONS

This simple process can be widespread in various fields related to the treatment of dehydrated fruits. It can be developed for the by-products often rejected by some food-agronomy industries.

Basis: ton of dehydrated prunes

Pollution balance:

	Rejects		Old Procedure	New Procedure
	throughput	m^3/t	1	0
Water	MES	kg/t	65	0
	COD	kg/t	80	0
	BOD	kg/t	50	0
	sorbic acid	kg/t	3	0
	waste solids in discharge		0	1

Economic balance: 1982 Dollars

	Old	New
Investment	—	111,340
Annual costs	—	90,620
Annual returns	0	257,240

OLD PROCEDURE

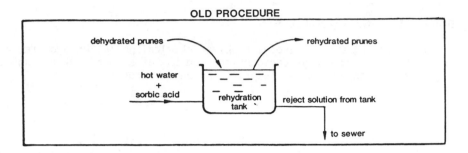

NEW PROCEDURE

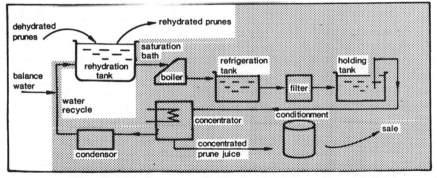

VEGETABLE PEELING (POTATO)
MECHANICAL PROCESSING AND RECYCLE

ADVANTAGES OF THE NEW PROCESS

This peeling process avoids risks of chemical pollution and reduces the clean water and energy consumption. On the other hand, it guarantees larger production output, while providing better products.

POSSIBLE EXTENSIONS

This process can be widespread only to vegetables of high consistency which eliminates all of those that are too fragile or too long. Fruits also could be mechanically peeled.

Basis: ton of treated potatoes

Pollution balance:

	Rejects		Old Procedure	New Procedure
	throughput	m^3/t	25	5
Water	MES	kg/t	25	12
	MO	kg/t	16	7
	Soda	kg/t	5	0
	citric acid	kg/t	6	1
Waste solids	peeling wastes	kg/t	150	150

Economic balance: 1982 Dollars

	Old	New
Investment	23,380	81,850
Annual costs	524,330	183,430

OLD PROCEDURE

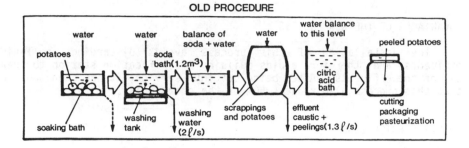

NEW PROCEDURE

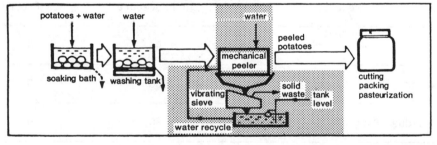

DISTILLERY
RECOVERY/REUSE AND RECYCLE OF EFFLUENT

ADVANTAGES OF THE NEW PROCESS

The installation of this process avoids discarding polluting effluents and therefore avoids building a purification station to treat all of the effluents. After being put into service, it has been noticed that this process also reduces 50% of the energy cost.

POSSIBLE EXTENSIONS

This process can be easily applied to most distillation factories.

Basis: m^3 distilled alcohol

Pollution balance:

	Rejects		Old Procedure	New Procedure
	throughput (red water)	t/m^3	44	0
Water	BOD$_5$	kg/m^3	1,500	0
Residue	BOD$_5$	kg/m^3	30	0

Economic balance: 1982 Dollars

	Old	New
Investment	–	705,300
Annual costs	–	–

OLD PROCEDURE

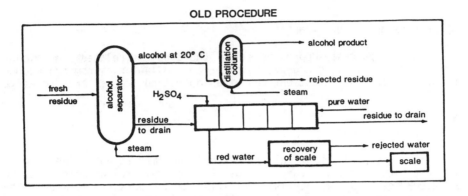

NEW PROCEDURE

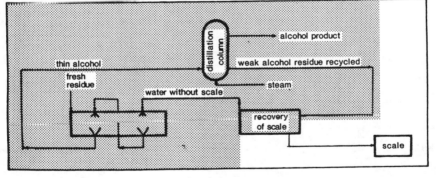

BREWERY
DRY RECOVERY OF KIESELGUHR

ADVANTAGES OF THE NEW PROCESS

The new process suppresses all of the problems related to clean-up using water, i.e., clogging, abrasion and pollution. From the economic point of view, the investment and the operating costs are relatively small.

POSSIBLE EXTENSIONS

The recouperation of malt beverages sometimes accepts the addition of kieselguhr muds. The use as an agricultural amendment has some difficulties (yeast autolysis), but it is always possible to store these pasty rejects, which is still better than cleaning up with water. The process of dry recovery of kieselguhr is very simple and should be extended to all breweries.

It can even be extended to the factories which filter wine with kieselguhr as "La Souete des vins de France de Lyons" does. This process tends to be widespread and the following list is not exhaustive.

Basis: 100 1(hl) filtrate

Pollution balance:

	Rejects		Old Procedure	New Procedure
	throughput	1/hl	8	6
Water	MES	g/hl	50	trace)
	BOD	g/hl	4	trace) rinse water
	COD	g/hl	8	trace)

Differential economic balance: 1982 Dollars

	Old	New
Investment (5 yr. liquidation for new)	-	35,750
Annual costs	-	4,200

OLD PROCEDURE

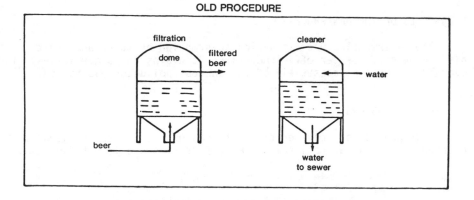

NEW PROCEDURE

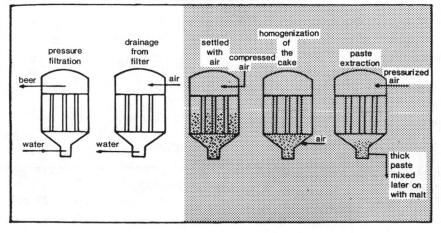

SUGAR REFINERIES
BY-PRODUCT RECOVERY BY CRYSTALLIZATION

ADVANTAGES OF THE NEW PROCESS

The regenerating process avoids consumption of water and salt. It also ensures a better efficiency since it avoids the sugar losses. The overall process suppresses all of the aqueous pollution discharged.

POSSIBLE EXTENSIONS

This process of regeneration of resins could advantageously be adapted to many sugar refineries. Many factories use it with success.

Basis: 1000 tons of juice (tons)

Pollution balance:

	Rejects		Old Procedure	New Procedure
Water	throughput	$m^3/1000t$	84	0
Waste-Water	MES	kg/1000t	150	0
	Chlorine ions	kg/1000t	950	0

Economic balance: 1982 Dollars

	Old	New
Investment	–	422,300
Annual costs	360,000	–
Annual returns	–	29,230

15

OLD PROCEDURE

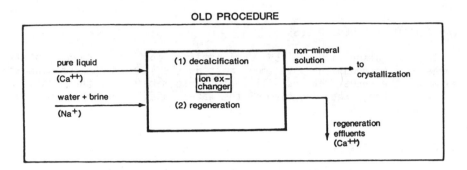

NEW PROCEDURE

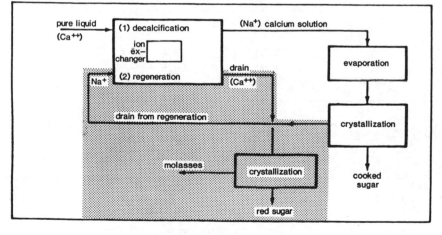

SUGAR PRODUCTION
RECOVERY OF DEMINERALIZATION EFFLUENT

ADVANTAGES OF THE NEW PROCESS

The pollution induced by demineralization has been totally suppressed. Although the investment cost is the same as that of standard treatment, the annual returns will increase to make up for that cost. On the other hand, the recycle of the water makes it possible to reduce water consumption by 50%.

POSSIBLE EXTENSIONS

All of the sugar refineries which demineralize the beet juice use a similar process. This process corresponds to the method which satisfies the requirement of the contract.

Basis: ton demineralized beet

Pollution balance:

	Rejects		Old Procedure	New Procedure
	throughput	m^3/t	0.75	0
Water	BOD	kg/t	4.5	0
	COD	kg/t	8.5	0
	salt	kg/t	20	0

Economic balance: 1982 Dollars

	Old	New
Investment	1,465,200	1,880,400
Annual costs	3.6	3.6
Annual returns	-	3.3

16

OLD PROCEDURE

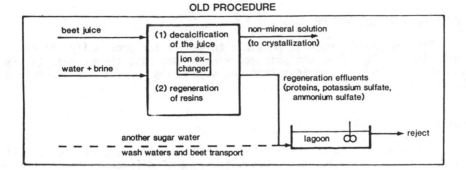

NEW PROCEDURE

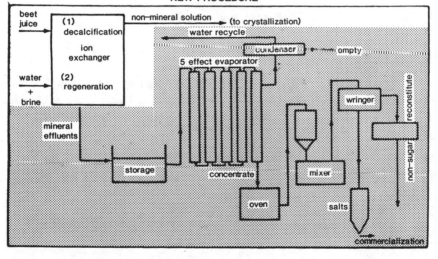

CANDY MANUFACTURING
RECYCLE TECHNIQUES

ADVANTAGES OF THE NEW PROCESS

The new process made it possible to reduce the previously discarded
pollution by 50%. From the point of view of savings, the process is also
interesting since the savings from the sugar is a profitable investment
by the end of the first year.

POSSIBLE EXTENSIONS

The amount of sugar which remains discarded into the effluents still
causes pollution problems. Because of the difficulty inherent in a
perceptible improvement of the recycled yield, consideration must be
given to the construction of a standard purification station in order to
treat the remaining discarded water.

Basis: ton of candy

Pollution balance:

	Rejects		Old Procedure	New Procedure
	throughput	m^3/t	3.5	3.5
Water	sugar	kg/t	35	19
	COD	kg/t	70	38
Solids	packing paper filter	g/t	–	850
	activated carbon	g/t	–	140
	lime	g/t	–	30
	soil filtrant	g/t	–	40

Economic balance: 1982 Dollars

	Old	New
Investment	–	37,800
Annual costs	48,700	9,880

17

OLD PROCEDURE

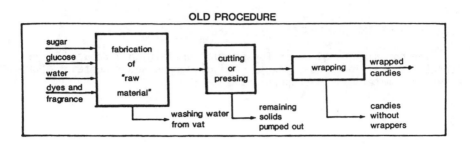

NEW PROCEDURE

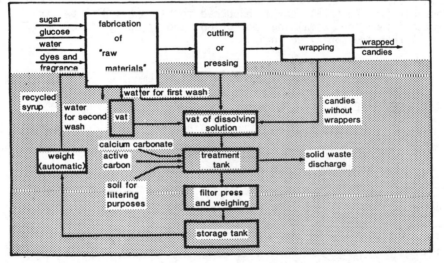

CONCRETE RELATED PRODUCTS
WATER CLEAN-UP AND RECYCLE

ADVANTAGES OF THE NEW PROCESS

The consumption of water for production is highly reduced by recycle (about one million m^3 per year). The discarded water contains about 300 times less materials in suspension.

POSSIBLE EXTENSIONS

The extension of the new process to other factories of the company is already planned. The total water recycle idea has some production problems created by the progressive concentration in salts, which can be avoided only by partially renewing the process water. There are hopes to solve these problems in 1983 and then proceed to a total recycle. The relatively small volume of actual pollution ($0.45m^3$/ton produced) could possibly (if the total recycle happened to be impossible) be treated by neutralizing the remaining pollution and ridding it of heavy metals (Cr^{+6}) that are contained in it.

basis: ton of asbestos-cement

Pollution balance:

	Rejects		Old Procedure	New Procedure
	throughput	m^3/t	5.7	0.45
Water	MES	g/t	2460	8.5
	MO	g/t	215	14

Economic balance: 1982 Dollars

	Old	New
Investment	–	1,607,750
Annual costs (differential)	–	146,160

18

OLD PROCEDURE

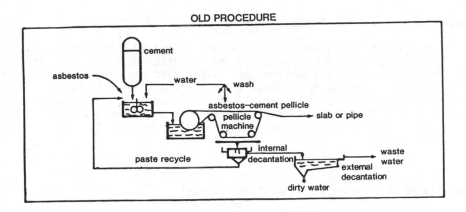

NEW PROCEDURE

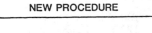

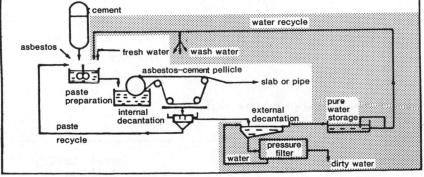

SAND AND STONE PRODUCTION
WASTEWATER RECYCLE

ADVANTAGES OF THE NEW PROCESS

The three solutions introduced (combined decanting basin, clarification, and dry separator) have different costs and different features in terms of the environment. These also correspond to different requirements. In any case, these process changes demonstrate that it is possible to wash stone-pit materials in a less polluting fashion and with reasonable cost. The problems that remain to be solved are those of the dehydration of the sludges (when water is used) and by the removal of discarded solids.

Investment Requirements (Decantation Process)

Company	Water Volume t/d	Overall Cost $ (1982)
1	100	15,870
2	120	15,212
3*	80	39,680

* costs elevated due to engineering design expenses

Investment Requirements (Clarifloculation Process)

Company	Water Volume t/d	Overall Cost $ (1975)
1	60	111,670
2	90	78,360
3	255	156,720
4	480	168,470

OLD PROCEDURE

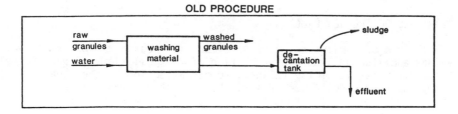

NEW PROCEDURE (DECANTATION)

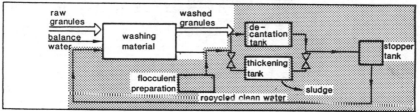

NEW PROCEDURE

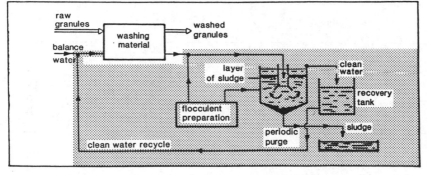

PETROLEUM REFINERY
AQUEOUS STREAM RECYCLE

ADVANTAGES OF THE NEW PROCESS

The use of various recycles have made it possible for some water savings and also an important reduction of the pollution. Direct recycle of the effluents leads the refinery to pay attention to the quality and finally to reduce pollution causes. Looking at the monetary savings, the recycle requires an important investment, while operating costs are about the same.

POSSIBLE EXTENSIONS

If this technique cannot be widespread, it still has an example value and could be applied to other refineries.

Basis: ton of refined raw petroleum

Pollution balance:

	Rejects		Old Procedure	New Procedure
	throughput	m^3/t	32	16
Water	MES	g/t	30	2.8
	BOD	g/t	60	3.6
	COD	g/t	125	17
	NH_4^+	g/t	50	15
	hydrocarbons	g/t	10	0
	phenols	g/t	2	0

Economic balance: 1982 Dollars

	Old	New
Investment	-	2,002,440
Annual costs	-	13,390

20

OLD PROCEDURE

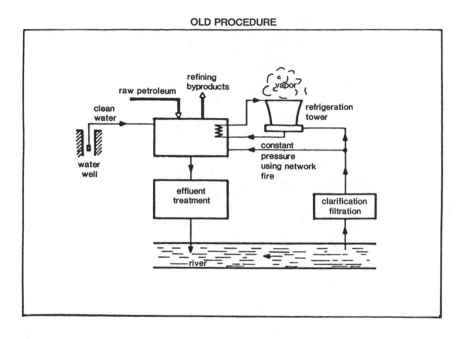

NEW PROCEDURE

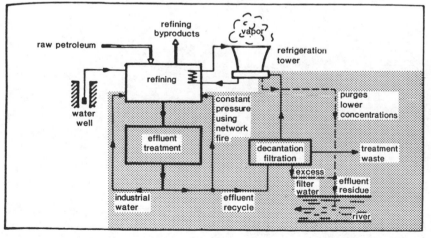

21

COKING FACILITIES
ALTERATION OF GASEOUS AND LIQUID DISCHARGES

ADVANTAGES OF THE NEW PROCESS

For coking, the new process eliminates liquid throughput. In addition, the air throughput does not have any more yellow coloring, which characterizes the presence of tar. The incineration of the smoke permits an important in-stream energy recovery.

For the carbonization, the liquid throughput is also eliminated. Eventually, the two processes lead to an important savings of the water consumption in the whole factory (about 60% of the total consumption).

Basis: ton treated wood

Pollution balance:

	Global Rejects		Old	New Coking	New Carbonization
	throughput	m^3/t	8.5	6.5	2.9
Water	MES	kg/t	1.2	0.5	0.2
	BOD	kg/t	5.6	0.8	0.5
	COD	kg/t	11.3	1.6	1
	MO	kg/t	7.5	1	0.6
Air			Visible yellow fumes	The yellow fumes indicate the presence of tar	

Economic balance: 1982 Dollars

	Old Coking	New Coking	Old Carbonization	New Carbonization
Investment	231,500	396,850	–	127,600
Annual costs	23,400	33,000	25,700	22,650

21 (Coking)

OLD PROCEDURE

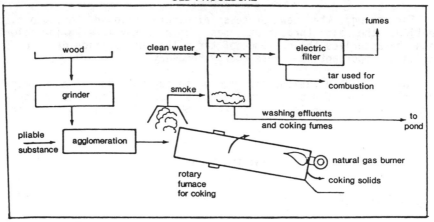

NEW PROCEDURE

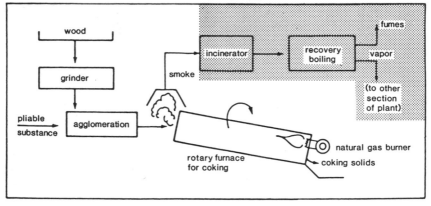

21 (Carbonization)

OLD PROCEDURE

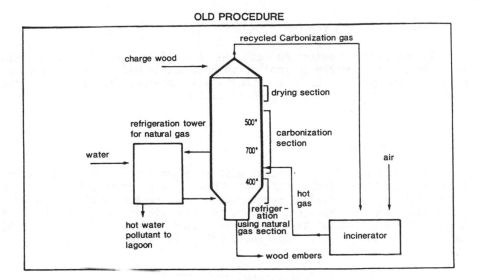

NEW PROCEDURE

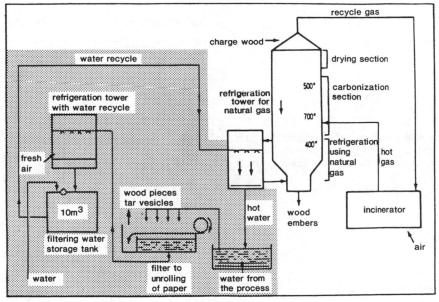

CHEMICAL MANUFACTURING (FURFURAL)
RECYCLE AND REUSE OF MATERIALS

ADVANTAGES OF THE NEW PROCESS

The new process described has made it possible, without a capital investment, to eliminate the pollution of about 14,000 kg/day of chemical oxygen demand (COD). Such a result was obtained by analyzing the production process, making some modifications, which finally proved to be not large in scope.

POSSIBLE EXTENSIONS

The extension of this process must be planned or engineered rather than simply transposed directly as a new process. Direct transposition is possible only in similar centers. The accurate analyses of the production process (process leading to reduction or even elimination of the pollution throughput by an internal recycle done at little expense) can be transposed in many different chemical factories.

Basis: ton of furfural

Pollution balance:

	Rejects		Old Procedure	New Procedure
Water	throughput	m^3/t	13.5	0
	COD	kg/t	540	–

The suppression of phosphoric juices being rejected from the system demands
high investments and depends on the precise production process.

22

OLD PROCEDURE

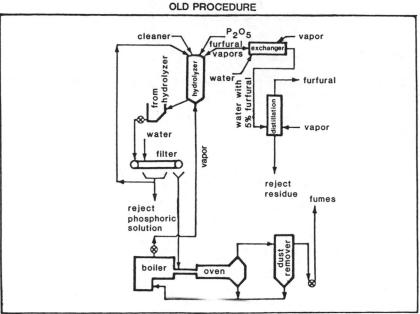

NEW PROCEDURE

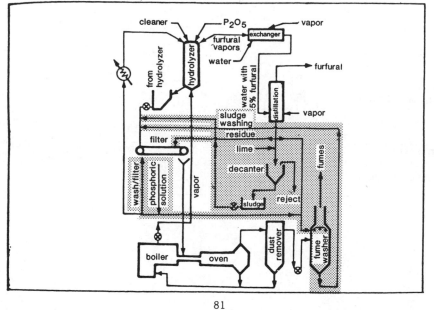

AMMONIA PRODUCTION
EXTRACTION FOR EFFLUENT CLEAN-UP

ADVANTAGES OF THE NEW PROCESS

Changing the process has suppressed the throughput of free ammonia to the river by a significant amount.

POSSIBLE EXTENSIONS

This recovery process is quite specific to the operating conditions of the gas clean-up unit for by-product by this factory. Hence it cannot be considered to extend it to other factories using a different cleaning-up unit.

Basis: ton of product

Pollution balance:

	Rejects		Old Procedure	New Procedure
	throughput	m^3/t	16.1	0
Water	ammonia	kg/t	2.85	-

Economic balance: 1982 Dollars

	Old *	New
Investment	-	568,000
Annual costs	-	295,000
Annual returns	-	268,900

* cannot estimate

OLD PROCEDURE

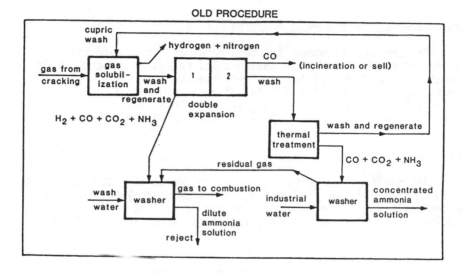

NEW PROCEDURE

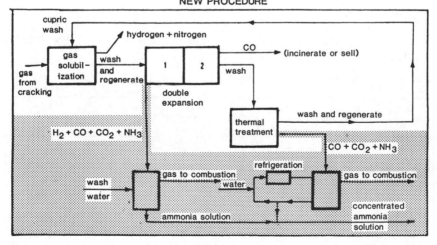

AMMONIA SYNTHESIS
HYDROGEN RECOVERY

ADVANTAGES OF THE NEW PROCESS

The ammonia throughput has been more or less suppressed for the whole factory. The purification using potassium carbonate seems to be beneficial for the environment. From the economic point of view, the operating costs per ton (10^3 Kg) of ammonia produced is reduced by 70%. This is mostly due to the energy savings. This shows the advantages of the new process in spite of the important capital investment.

POSSIBLE EXTENSIONS

The purification using the potassium carbonate already exists in many units and can be applied to any ammonia-synthesis factory. The economical interest is reinforced by the fact that it produces energy savings for the system and therefore there is possible broad extension.

Basis: ton of ammonia product

Pollution balance:

	Rejects		Old Procedure	New Procedure
Water	Ammonia	E/t	490	10
		kg/t	3.3	0.03
Air	Ammonia	kg/t	0.03	0

Economic balance: 1982 Dollars

	Old	New
Investment	771,700	4,677,100
Annual cost	973,422	140,300
Annual returns	122,800	0

OLD PROCEDURE

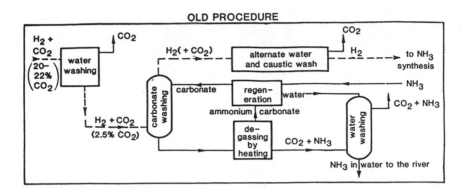

NEW PROCEDURE

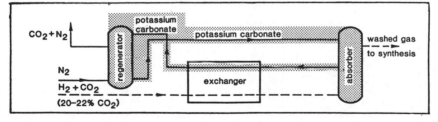

25

PHOSPHORIC ACID PRODUCTION
STREAM SEGREGATION AND RECYCLE

ADVANTAGES OF THE NEW PROCESS

This process modification permits an important reduction in pollution of fluorine ions. The new process produced a small reduction in water consumption, either directly by recycle or indirectly by discharging a small amount of polluted fabrication water downstream. The process has required no specific equipment and only a few changes in the mainstream lines.

POSSIBLE EXTENSIONS IN THE FUTURE

The particular separation of the water can be widespread according to various activities. It makes the treatment of effluents easier and in some cases allows particular recycling.

Basis: ton phosphoric acid produced

Pollution balance:

	Rejects		Old Procedure	New Procedure
	throughput	m^3/t	20	15
Water	fluid ions	kg/t	5	0.5

Economic balance: 1982 Dollars

	Old	New
Investment	–	122,100
Annual costs	42,846	1,928

25A

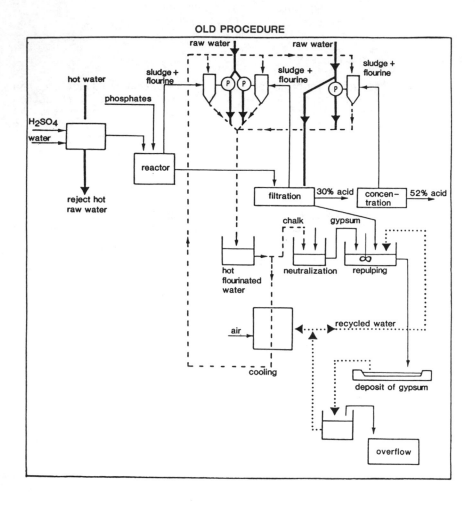

88

25B

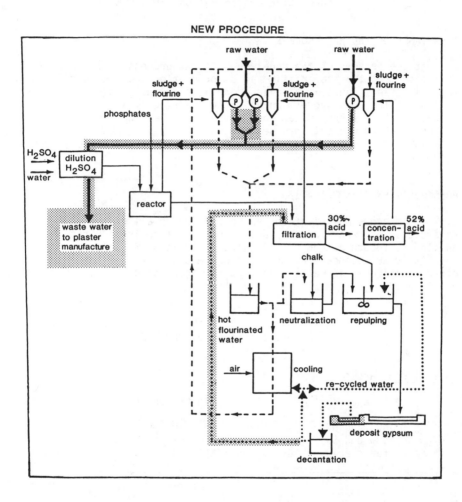

NEW PROCEDURE

89

NITRIC ACID PRODUCTION
SUCCESSIVE CONCENTRATION TECHNIQUE

ADVANTAGES OF THE NEW PROCESS

At the factory of "La Madeleine", a new process has decreased the pollution related to nitric acid production and the concentration of sulfuric acid. There is also an economical advantage due to the energy savings. The new process makes the processing cycle easier because the reconcentration workroom can be removed. The workroom for the new process, which was put into service in late 1979, might still be changed and improved. The process itself and/or the technology could be improved.

POSSIBILITIES OF APPLICATION

The process modification can be applied to all of the new factories regardless of what nitric acid concentration is required. It can be applied to modernization of old units or it is profitable for increasing the nitric acid concentration. The process is not profitable compared to other techniques, if we only look at the decrease of pollution in gaseous nitric acid.

Basis: ton of acid produced

Pollution balance:

	Rejects		Old Procedure	New Procedure
Air	NO_x and HNO_3	kg/t	10	4.5
	SO_2 and Sulfuric acid	as kg H_2SO_4/t	3.5	none

Economic balance: 1982 Dollars

	Old	New
Investment	–	15,397,800
Annual costs	–	95
Annual returns	–	21

26

OLD PROCEDURE

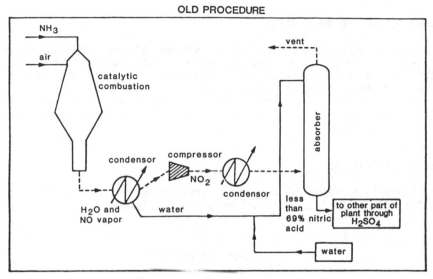

NEW PROCEDURE

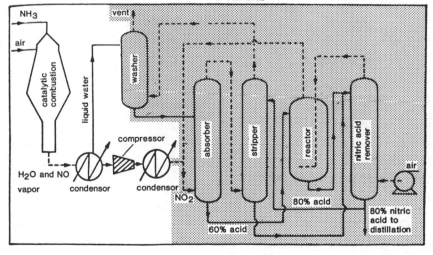

FERTILIZER MANUFACTURING
GAS CONDENSATION FOR ENHANCED PRODUCT RECOVERY

ADVANTAGES OF THE NEW PROCESS

The scrubbing of the evaporation stream of the ammonium nitrate substantially reduces ammonia pollution, both in the air and in the water. For a small expenditure (one device and some piping), ammonia is saved and this makes up for the initial expense. The consumption of energy is important in the factory. The overall energy balance (non-consumption of energy needed to supply 1 ton (1000 kg) of ammonia per day is positive.

POSSIBLE EXTENSIONS

There is no technical problem to install such a scrubbing device, and the economical interest might help its development.

Basis: ton of nitrate

Pollution balance:

	Rejects		Old Procedure	New Procedure
	throughput	m^3/t	0.27	0.27
Water	nitrogen material (ammonia or nitrate)	kg/t	1.6	0.4
Air	nitrogen	kg/t	1.5	0.35

Economic balance: 1982 Dollars

	Old	New
Investment	–	115,750
Annual costs	96,460	12,860

OLD PROCEDURE

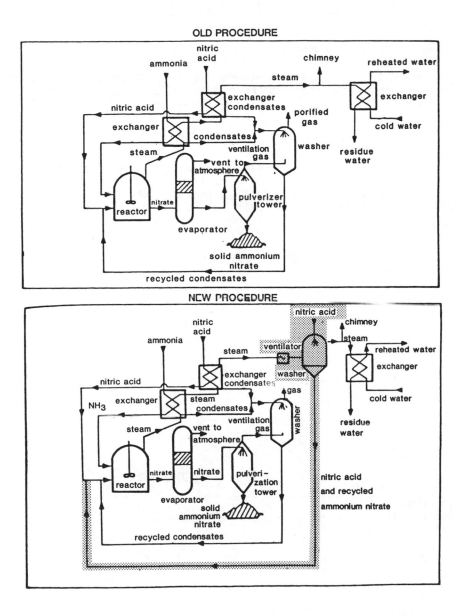

NEW PROCEDURE

FERTILIZER MANUFACTURING
PROCESS CONTROL AND SEPARATION FOR ENHANCED PRODUCT RECOVERY

ADVANTAGES OF THE NEW PROCESS

In addition to the important reduction of pollution, the new process leads to a slight rise in production which quickly balances the investment.

POSSIBLE EXTENSIONS

The extension of this process can be done at other ammonium nitrate production units with the same production devices. Another production concern is for a better stoichiometric control (the installation of a solid removal system requires special operating conditions).

Basis: ton of product

Pollution balance:

	Rejects		Old Procedure	New Procedure
	throughput	l/t	450	450
Water	ammonia	g/t	3,700	280
	nitrate	g/t	3,500	160

Economic balance: 1982 Dollars

	Old	New
Investment	–	152,000
Annual costs	same	same
Annual returns	–	292,300

28

OLD PROCEDURE

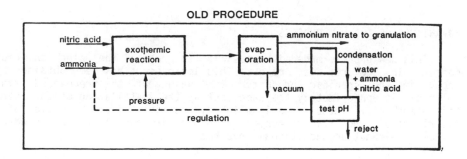

NEW PROCEDURE

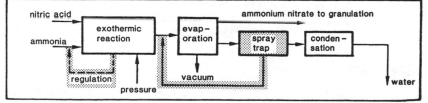

ZINC SLUDGE RECOVERY
CAKE WASHING AND EFFLUENT REDUCTION

ADVANTAGES OF THE NEW PROCESS

This new process made it possible to reduce the zinc chloride content in the residual cakes. This improvement has been obtained by recovering the specific product. The value of the recovered zinc chloride is approximately the same cost as the installation of the extra filtration unit. The balance is slightly positive. The realization of a watertight pit and effluent recycle have made it possible to reduce water pollution without any additional investment.

POSSIBLE EXTENSIONS AND IMPROVEMENTS

The step consisting of taking the residual cakes and extracting from them the polluting material by washing or filtration can be widespread to a large number of different chemical companies. However, this type of operation requires a certain concentration of the remaining filtrate.

Basis: none

Pollution balance:

Rejects		Old Procedure	New Procedure
zinc chloride	kg/t	10	1
coke		–	+10%
plant effluents	$Zn(mg/m^3)$	200	20

Economic balance: 1982 Dollars

	Old	New
Investment	–	244,700
Annual costs	2,923	87,700
Annual returns	–	350,000

OLD PROCEDURE

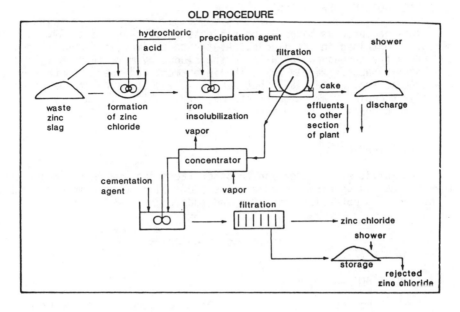

NEW PROCEDURE

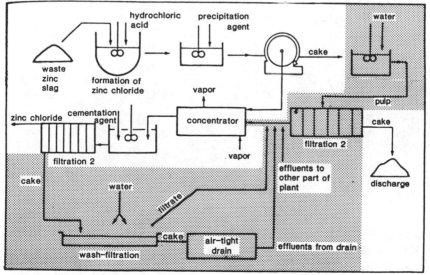

SLUDGE PROCESSING
MERCURY REMOVAL

ADVANTAGES OF THE NEW PROCESS

This new process makes it possible to recover at least 99% of the mercury contained in the sludge. At the same time, it makes it possible to reduce the throughput of NaCl in the sludge by more than 80% and in the water by approximately 70%. The investment and operating cost of the new process are substantial, but they were necessary to succeed in greatly reducing the amount of mercury, which is a very dangerous polluting product, rejected into the water.

POSSIBLE EXTENSIONS

Possibilities of extension to other factories are rather restricted. The process is specific to electrolysis of chlorine on a mercury cathode producing both potassium hydroxide and soda.

Basis: ton of chlorine

Pollution balance:

	Rejects		Old Procedure	New Procedure
	throughput	m^3/t	3.75	3.75
	MES	g/t	3,600	3,600
	BOD	g/t	0	0
		Eq/t	900	30
	Mercury	g/t	2.5	0.07
	NaCl	g/t	145,000	50,000
Waste Solids (sludge)	Mercury	g/t	100	0.5
	NaCl	g/t	30,000	500

Economic balance: 1982 Dollars

	Old	New
Investment	-	3,507,800
Annual costs	138,270	391,700

OLD PROCEDURE

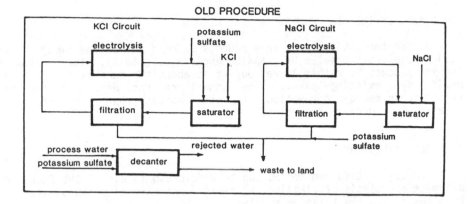

NEW PROCEDURE

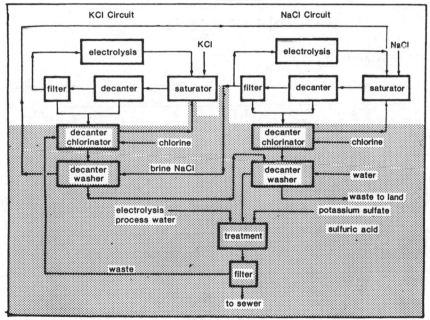

ELECTROCHEMICAL PRODUCTION OF SODIUM CHLORATE
ELECTRODE PROCESS MODIFICATION

ADVANTAGES OF THE NEW PROCESS

As for the environment, this process reduces all of the sludge that would have been created by graphite anode electrolysis. The investment for the process is rather large, but it is about the same as the cost to modify the existing plant. The annual returns, especially from the reduction in energy consumption, are high enough that the return on investment is done in two years.

POSSIBLE EXTENSIONS

The use of titanium anodes can be widespread to all of the factories using electrolysis of alkaline chlorides. The extension could have some problems due to the titanium supply.

Basis: ton of $NaClO_3$

Pollution balance:

	Rejects		Old Procedure	New Procedure
Waste Solids (sludge)	graphite dust in sludge form	kg/t	6	nil

Economic balance: 1982 Dollars

	Old	New
Investment	-	2,734,370
Annual costs	4,512,200	3,186,670

OLD PROCEDURE

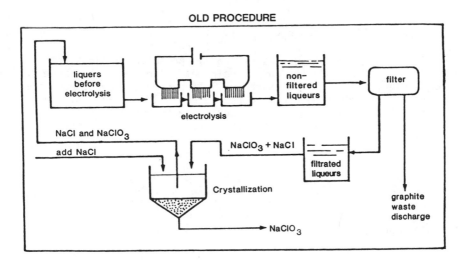

NEW PROCEDURE

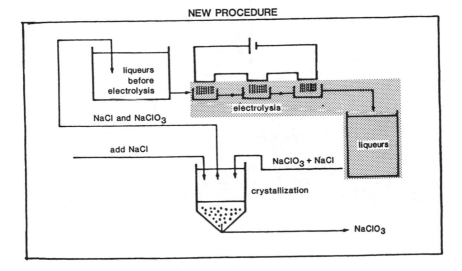

32

HYDRAZINE HYDRATE MANUFACTURING
NEW PROCESS SUBSTITUTION

ADVANTAGES OF THE NEW PROCESS

The new process suppresses all polluting throughput. The distillation by-products are incinerated and sent to the cement manufacturer. It also reduces the pollution created downstream where the by-products of hydrazine are made from a non-salted solution. There is a significant decrease in production costs due to the energy savings. The extra investment necessary is small considering the benefits of the process.

Basis: ton hydrazine hydrate

Pollution balance:

	Rejects		Old Procedure	New Procedure
	throughput	m^3/t	27	0
Water	COD	kg/t	29	–
	Chlorine ions	t/t	2.4	–
	ammonia	kg/t	2.4	–
	hydrozine	kg/t	1.3	–
Air	ammonia	kg/t	5.3	–

Economic balance: 1982 Dollars

	Old	New
Investment	13,154,350	15,2000,600
Annual costs	–	40% reduction

102

OLD PROCEDURE

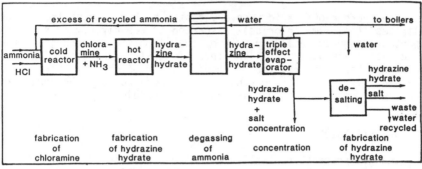

NEW PROCEDURE

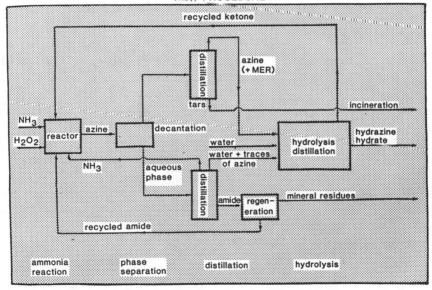

CHLORINATED RESIDUE TREATMENT
ENHANCED RECOVERY OF HYDROCHLORIC ACID

ADVANTAGES OF THE NEW PROCESS

The hydrochloric acid recovery system highly reduces the amount of chlorides (calcium chloride) discarded to the river. It suppresses an important dangerous pollution. From the economic standpoint, the new installation requires a high investment (approximately 50% more than would have been required for the modification of the existing process). The small lime consumption reduces the annual costs. The annual returns from the hydrochloric acid sales make it possible for the new process to pay for itself.

POSSIBLE EXTENSIONS

The new process, now well controlled, treats 97% of chlorinated remains of St. Auban Factory and a part of chloe-chimie factories. It can be widespread to some factories; 6 permits have already been delivered (Spain, Morocco, USSR) and others are going to be delivered soon.

Basis: ton of treated chlorine residues

Pollution balance:

	Rejects		Old Procedure	New Procedure
	throughput:			
	to the neutralizer	m^3/t	1.5	1.5
	to the quench	m^3/t	66.5	0
Water	Total:		68	
	toxin	kE/t	2	E
	–	kg/t	670	55
	(calcium and soda)			

Economic balance: 1982 Dollars

	Old	New
Investment	2,923,190	4,384,780
Annual costs	1,381,200	982,190
Annual returns	–	1,315,430

OLD PROCEDURE

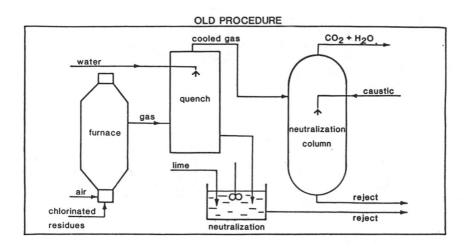

NEW PROCEDURE

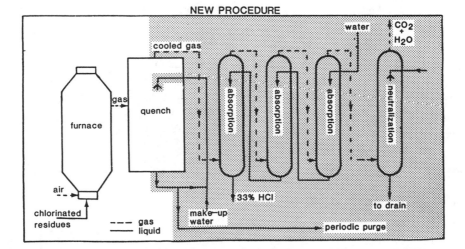

CHEMICAL MANUFACTURING (STYRENE)
RESIDUE RECOVERY OF PRODUCT

ADVANTAGES OF THE NEW PROCESS

This dry neutralization process with recovery of salt, suppresses all aqueous wastes that contain hydrocarbons, oils, and salts. It reduces water consumption and simplifies the neutralization steps by not greatly increasing the output.

POSSIBLE EXTENSIONS

Such a process can profitably be applied to the treatment of solutions containing acid catalyst. One factory has a patent on the process and plans to sell it. It combines a modern non-polluting technique (dry neutralization) and a standard process of making ethylbenzene, using aluminum trichloride catalysis.

Basis: ton of ethylbenzene produced

Pollution balance:

	Rejects		Old Procedure	New Procedure
	throughput	m^3/t	1.5	0
Water	MES	kg/t	2	0
	MO	kg/t	3	0
	toxin	Eq/t	0.003	0
	Waste solids	kg/t	0	9

Economic balance: 1982 Dollars

	Old	New
Investment	1,215,300	1,590,950

OLD PROCEDURE

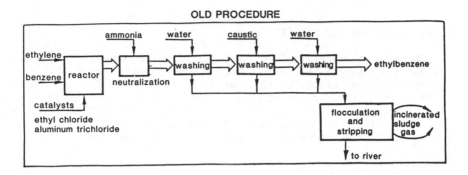

NEW PROCEDURE

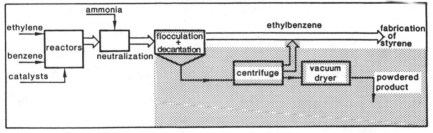

POLYMER PRODUCTION (PVC)
VINYL CHLORIDE MONOMER RECOVERY

ADVANTAGES OF THE NEW PROCESS

The noxious releases of VCM (vinyl chloride monomer) are almost totally suppressed. The resins contain a very small amount of VCM (from 1 to 2 ppm instead of 800 to 1000) at the moment when they are released to the atmosphere, prior to drying. From the economic standpoint, the basic investment is very high and the recovery of VCM does not allow a very rapid return on investment. In terms of working costs, there is an expense due to the consumption of 200 kg of vapors per ton $(10^3$ kg) of PVC.

POSSIBLE EXTENSIONS

Very good results have led "NTO-Chimie" to build a second PVC factory using the same process in Balan. The process can be used in all of the factories producing PVC by polymerization of VCM suspension.

Basis: ton of PVC

Pollution balance:

	Rejects		Old Procedure	New Procedure
Air	VCM vapor	kg/t	6.3	0.025

Economic balance: 1982 Dollars

	Old	New
Investment (differential)	-	1,753,900
Annual costs	-	22,200

OLD PROCEDURE

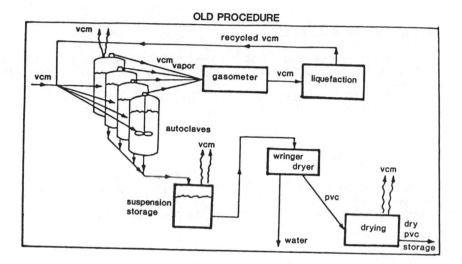

NEW PROCEDURE

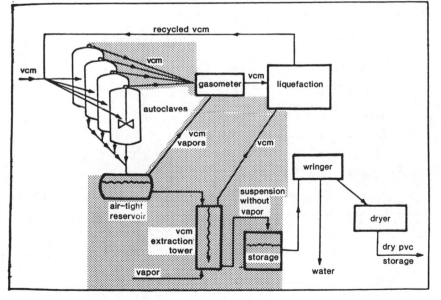

CHLORAL PRODUCTION
SEPARATION FOR IMPROVED RECOVERY

ADVANTAGES OF THE NEW PROCESS

The former process had a tremendous waste throughput. The old process required the use of very concentrated, hot sulfuric acid and that is a health hazard for the workers. Putting the new process into service made it possible to completely eliminate the throughput and to improve security conditions of the workplace by replacing the acid with a solvent in a closed network.

POSSIBLE EXTENSIONS

The principle of dehydration with a solvent can be widespread to other processes that normally use sulfuric acid. Every case must be examined separately in order to choose the necessary solvent.

Basis: ton of hydrates

Pollution balance:

	Rejects		Old Procedure	New Procedure
	throughput	l/t	360	140
Water	sulfuric acid	kg/t	600	0
	chlorine derivatives	kg/t	25	0
	COD	kg/t	6	20

Economic balance: 1982 Dollars

	Old	New
Investment	not estim.	3,308,800
Annual costs	12,832,800	12,482,000

36

OLD PROCEDURE

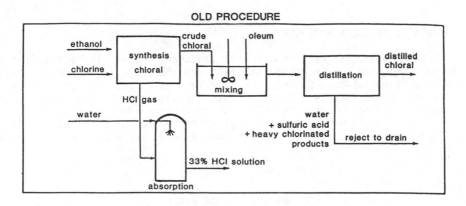

NEW PROCEDURE

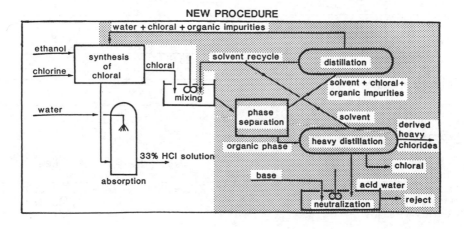

111

ORGANIC SYNTHESIS REACTORS
CATALYST CHANGE AND HEAT EXCHANGE IMPROVEMENT

ADVANTAGES OF THE NEW PROCESS

The new process has two peculiar features. First, it is possible to avoid using soda as a reaction catalyst and therefore it suppresses major amounts of pollution. Second, it uses a heat transfer process that recycles approximately 80% of the water supplied at the beginning of the process. The new process is efficient in controlling pollution from the economic standpoint. Indeed the yield of the reaction is less in terms of the new process, but it also consumes energy to heat the secondary circuit.

Basis: ton of ester

Pollution balance:

	Rejects		Old Procedure	New Procedure
	throughput	m^3/t	0.7	0
Water	COD	kg/t	26.5	–
Air	NO/NO_2		0	–

Economic balance: 1982 Dollars

	Old	New
Investment	2,359,200	4,713,000
Annual costs		2,700,370
Annual returns		1,881,200

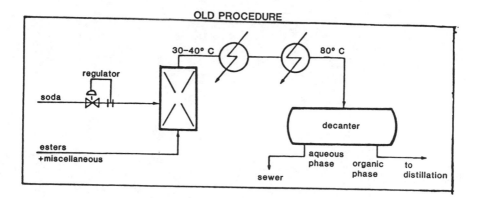

OLD PROCEDURE

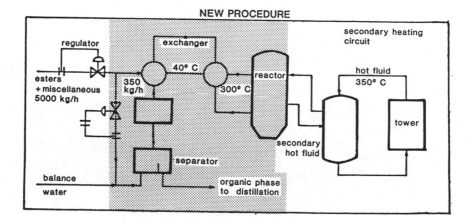

NEW PROCEDURE

38

DETERGENT MANUFACTURING
EFFLUENT RECYCLE

ADVANTAGES OF THE NEW PROCESS

The effluent recycle in the new process greatly reduces the pollution generation. From an economic point of view, the savings on water and raw materials is small in comparison to the operating costs. But as a matter of fact, the factory could not have kept on polluting as much as it was. The recycling in the new process reduces the pollution to a greater degree than that of a purifying station, especially in terms of oxidizable materials. The investment and operating costs are less than that of a purification station.

Basis: none

Pollution balance:

	Rejects		Old Procedure	New Procedure
	MES	kg/hr	810	150
Water	MO	kg/hr	850	100

Economic balance: 1982 Dollars

	Old (1)	New (2)
Investment	1,519,000	703,340
Annual costs	212,670	124,570

(1) cost estimation for the biological purification system
(2) equipment for recycling the effluents

38A

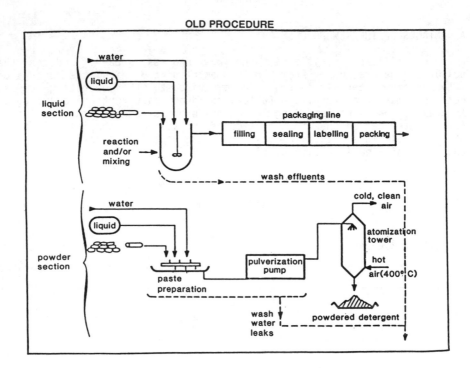

38B

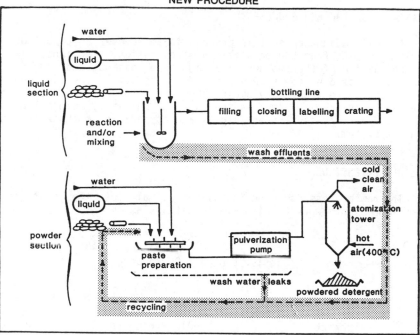

39

ZINC ORE REFINING
AIR EMISSION RECOVERY

ADVANTAGES OF THE NEW PROCESS

The main advantage of this process is to recover 80 to 90% of sulfur dioxide escaping from the catalytic neutralization. The innovation is in the combination of the two purification steps (gaseous and liquid) which makes it possible to save up to 30% of the lime. The lime would have been necessary to separately purify the two throughput streams and it makes it possible to improve the catalyst output by reinjecting sulfur dioxide. Thus the conversion outputs are comparable to those of a double catalyst system.

POSSIBLE EXTENSIONS

The new process could be installed at all of the sulfuric acid production plants working with simple catalysis.

Basis: ton of ZnO produced

Pollution balance:

	Rejects		Old Procedure	New Procedure
Air	SO_2	kg/t	22	3
Waste Solids	$CaSO_2$ (from sludge)	kg/t	30	42

Economic balance: 1982 Dollars

	Old	New
Investment	321,340	803,360
Annual costs	133,890	174,000
Annual returns	–	40,170

OLD PROCEDURE

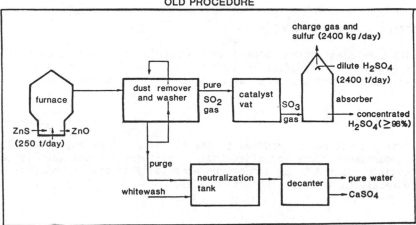

NEW PROCEDURE

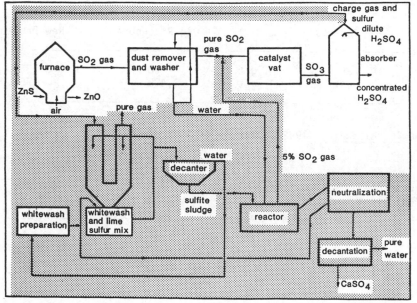

COMBUSTION OVENS
IMPROVED SCRUBBING TECHNIQUE

ADVANTAGES OF THE NEW PROCESS

The scrubbing of the smoke from the burning oven uses a part of the coke required to manufacture anodes. The operation is well integrated with the manufacturing process without generating secondary waste.

POSSIBLE EXTENSIONS

This process can be applied to the purification of any smoke loaded with condensable tars provided that it is correctly set up. It can be particularly applied to an anode oven especially if the tar materials are used directly to manufacture these anodes.

Basis: ton of precooked anodes

Pollution balance:

	Rejects		Old Procedure	New Procedure
Air	tars	g/t	750	35
	dust	g/t	3,400	175

Economic balance: 1982 Dollars

	Old	New
Investment		668,400
Annual costs		18,740

40

OLD PROCEDURE

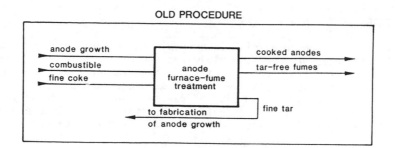

NEW PROCEDURE

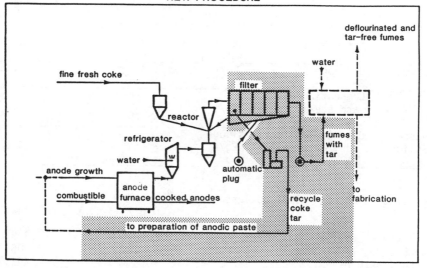

METAL CLEANING (STEEL)
SLUDGE REDUCTION AND METAL RECOVERY

ADVANTAGES OF THE NEW PROCESS

In the environmental field, the new process considerably reduces the amount of sludge produced and minimizes the amount of acid throughput to neutralize. From the economic point of view, the initial investment is very important. But, in spite of increased energy expenses, the savings from the reduction in consumption of raw materials considerably reduces the operating costs. The initial investment is profitable due to the reduction in operating costs and by the sale of ferric oxides.

POSSIBLE EXTENSIONS

The process described above could be used for a cold laminating conveyor equipped with hydrochloric cleaning and for the recovery of metals in solution. As far as hot cleaning is concerned, many units already use this process.

Basis: ton of etched steel

Pollution balance:

	Rejects		Old Procedure	New Procedure
	throughput	m^3/t	0.3	0.2
Water	Cl ions	kg/t	1.3	0.25
Air	Combustion gas and steam	m^3/t	–	45
Waste Solids	Decantation ferric sludge (dry material) kg/t		7	0.6

Economic balance: 1982 Dollars

	Old	New
Investment	1,110,800	7,308,000
Annual costs	2,657,180	1,306,700
Annual returns	0	292,300

OLD PROCEDURE

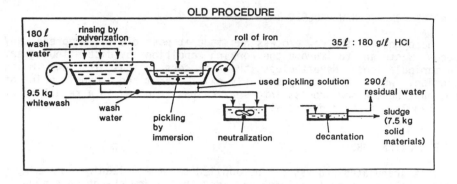

NEW PROCEDURE

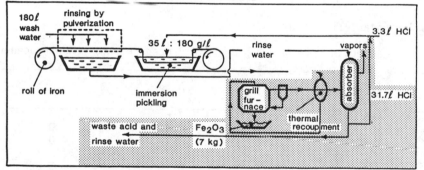

METAL CASTING CLEANING
DRY PROCESS SUBSTITUTION

ADVANTAGES OF THE NEW PROCESS

The change to the new process has made it possible to suppress any aqueous waste and to improve the working conditions by discontinuing any acid manipulation. Nevertheless this process brings its own noise pollution, which did not previously exist. From the economic point of view, putting the new process into service has resulted in a reduction of more than 35% of the annual costs.

POSSIBLE EXTENSIONS

This granulated process can be applied to the cleaning of simple shaped metallic pieces (castings) provided that a few changes are made.

Basis: ton of etched steel

Pollution balance:

	Rejects		Old Procedure	New Procedure
	throughput	m^3/t	0.18	0
Water	MES	kg/t	7.6	0
	MO	kg/t	0.4	0
	toxin	kg/t	0.1	0
Air	dust	kg/t	0	0.022
	vapor	kg/t	water & acid	0
Noise		decibles	-	80

Economic balance: 1982 Dollars

	Old	New
Investment	30,380	164,970
Annual cost	115,450	72,900

OLD PROCEDURE

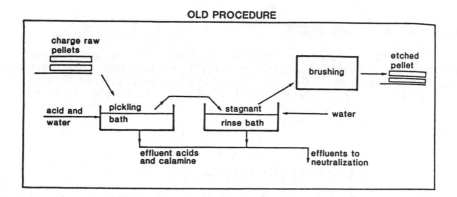

NEW PROCEDURE

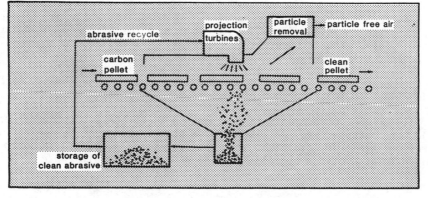

METAL CLEANING (STEEL)
RECYCLE AND RECOVERY TECHNIQUES

ADVANTAGES OF THE NEW PROCESS

The classical method of treatment of lime pickling does not make it possible to eliminate the nitrate ion and requires huge baths that must flow into a registered site. Up to now the new process used here is the most reliable one for treating pickling baths. The initial investment is very high and reflects the technical difficulties of putting the incineration unit into service.

POSSIBILITIES OF IMPROVEMENT AND EXTENSION

Several possibilities of improvement of the current process are as follows:

1) Treatment of the remaining rinse by:

    a) neutralizing with lime;
    b) mechanically dehydrating the sludges;
    c) adding the sludges to used baths before incineration.

2) A better evaluation of the condition of the baths while used in order to make the decision of when to reject used baths based on an objective criteria rather than a subjective criteria. This could make it possible to increase the longevity of these baths.

3) Regeneration of the acid contained in the used baths before incineration (but for the moment this is not reliable).

All of the improvements carried out on the process at one factory could be adapted to all steps of stainless steel scouring with hydroflourin/Nitric acid baths and maybe for some steps this could be applied to other types of acid scouring of metals and alloys.

Pollution balance:

	Rejects		Old Procedure	New Procedure
	throughput	m^3/t	53.95	7.7
Rinse Water	F^- ions	kg/t	0.85	0.5
	NO_3^- ions	kg/t	1.15	0.65
	iron	kg/t	1.9	1.1
	throughput	m^3/t	0.05	0.025
Used Juices	F^- ions	kg/t	0.4	0.75
	NO_3^- ions	kg/t	0.6	1.1
	iron	kg/t	1	1.8

Economic balance: 1982 Dollars

	Old	New
Investment	*14,616,000	3,069,150
Annual costs (differential)	–	3
Annual returns	–	.73

* estimated

OLD PROCEDURE

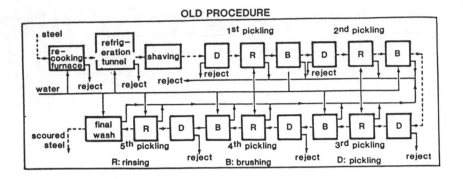

NEW PROCEDURE

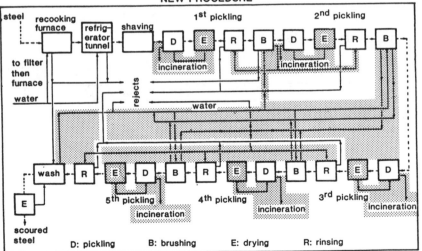

METAL CLEANING (BRASS)
SOLVENT SUBSTITUTION

ADVANTAGES OF THE NEW PROCESS

Due to this new process, polluted effluent is almost completely suppressed. The acid baths must be sent (in limited volume) to the detoxification center, then to filtration and monitoring the rinse water which is discharged. The inherent risks due to the use of nitric acid (in high concentration) are eliminated. From the economical standpoint, the decrease of expenses for materials is much greater than the increase of energy consumption. Hence, operating costs are lower for the new process and the initial investment is half of the one that would be necessary for a treatment installation.

POSSIBLE EXTENSIONS AND IMPROVEMENTS

The new process shows surface defects while the old one used to unify the top surface of the parts. Thus the new process requires special care for the moulding. Commercially, the parts may not look perfect, and hence slows down the possibilities of extension. The double interest of tribofinishing (economic savings and environmental concerns should make it possible for the new process to be widespread).

Basis: ton of treated pieces

Pollution balance:

	Rejects		Old Procedure	New Procedure
	throughput	m^3/t	40	2
Water	Acid	g/t	pure nitric acid	weak acid
	Cu	g/t	8,000	75
	Zn	g/t	5,500	450
Air	nitric vapors		30-40% of acid used	nil

Economic balance: 1982 Dollars

	Old	New
Investment	150,800	77,460
Annual costs	38,600*	19,300

*Estimation

44

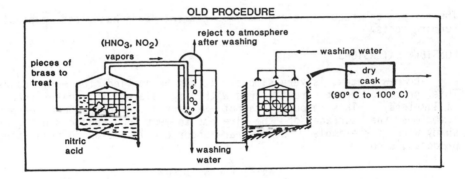

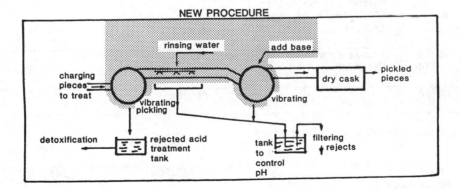

METAL DECARBONIZATION
DRY PROCESS SUBSTITUTION

ADVANTAGES OF THE NEW PROCESS

The new process makes it possible to be free of all constraints due to the utilization of scouring in acid solutions; that is:

- An installation that resists acid solution attacks.
- An installation for the treatment of used acid solutions, rinsing waters with an evacuation problem and storage of neutralization needs.
- Means of heating acid solutions.
- Waste and vapors from polluting acid inside and outside of the factory.

The new process requires a smaller investment and has less expensive running costs.

POSSIBLE EXTENSIONS

The new process is very simple. It can be applied to many cases. This process cannot treat wires with large diameters (greater than 25 millimeters). This process cannot be installed everywhere because sometimes the surface of the wire has to meet some requirements (very shiny wire, for example). Studies are under consideration to treat other products, also.

Basis: ton of wire

Pollution balance:

	Rejects		Old Proc.	New Proc.
Air	Volume air pollution (from the acid vapor) treated	m^3/t	10,000	-
Waste Solids	Sludge from neutralization	kg/t	55	-
	Calamine	kg/t	-	5
Noise		decibels	85	70

Economic balance: 1982 Dollars

	Old	New
Investment	730,800	146,150
Annual costs	15	7

OLD PROCEDURE

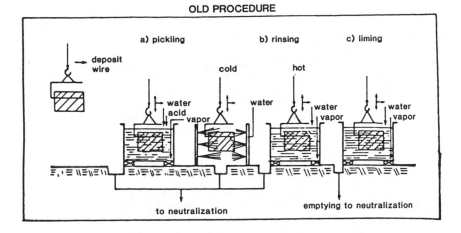

NEW PROCEDURE

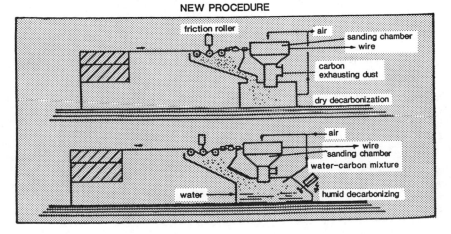

METAL ETCHING
PROCESS CHEMISTRY CHANGE

ADVANTAGES OF THE NEW PROCESS

The alcohol etching process suppresses all of the polluting wastes. The inherent risk of using sulfuric acid is also eliminated. Economically, the cost of using alcohol with this installation for etching is neglected. The more important operating costs are balanced by the amortization of the purification station that should have been installed and of the insurance costs implied.

POSSIBLE EXTENSIONS

The new process can be applied only to installations of casting and continuous laminating. The alcohol etching is viable only when superficial oxidation is light and when the copper is maintained to a high enough temperature in order for the chemical reaction to occur.

Basis: ton of wire machined

Pollution balance:

	Rejects		Old Procedure	New Procedure
Water before neutral- ization	throughput	m^3/t	0.75	0
	MES	g/t	20	–
	Toxin	Eq/t	>0.300	–
	MO	g/t	15	–

Economic balance: 1982 Dollars

	Old	New
Investment	877,000	–
Annual costs	90,600	222,160

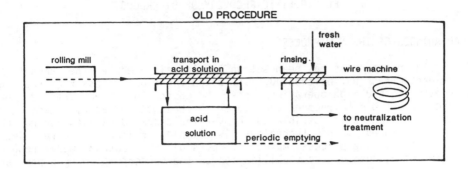

OLD PROCEDURE

rolling mill

transport in acid solution

rinsing

fresh water

wire machine

acid solution

periodic emptying

to neutralization treatment

NEW PROCEDURE

rolling mill

transport in alcohol solution

wire machine

alcohol for tank level

alcohol solution

METAL PROCESSING (COPPER)
ELECTROLYTIC RECOVERY OF BY-PRODUCT

ADVANTAGES OF THE NEW PROCESS

The new process has been set up to largely reduce the amount of effluents that must be treated. The detoxification process requires high investments and high operating costs. Moreover, the treatments do not make it possible to recover the copper which is stored as hydroxide. The new process makes it possible to recover the copper dissolved during the metal cleaning. The technology is simple, reliable and relatively inexpensive. The process makes it possible to obtain more regular cleanings because of the stabilization of the concentration of copper and acid in the solution.

POSSIBLE EXTENSIONS

All of the units producing copper using the transportation at high temperature can use this new process if they have to perform metal cleaning.

Basis: ton treated pieces

Pollution balance:

	Rejects		Old Procedure	New Procedure
Detoxif-ication station	copper sulfate free sulfuric acid (saturated)	g/t g/t	250 500	40 50

Economic balance: 1980 Dollars

	Old	New
Investment	-	42,000
Annual costs	-	1,890
Annual returns	-	5,400

OLD PROCEDURE

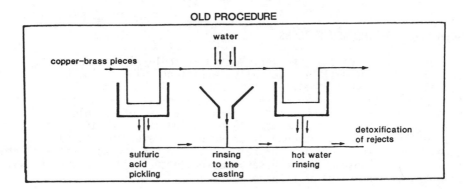

NEW PROCEDURE

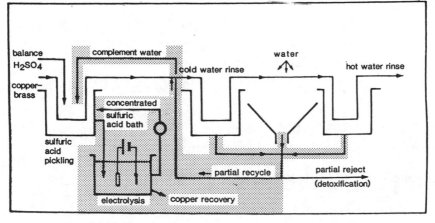

METAL OXIDE RECOVERY
FILTRATION/RECYCLE

ADVANTAGES OF THE NEW PROCESS

The process of dust recovery, using a handle-filter, which has been set up at Montreuil is actually efficient since almost all of the pollution previously emitted is suppressed. From the economic standpoint, the value of the recovered metals makes it possible to recover the investment within two years. The use of handle-filters is now very widespread in the metallurgical industry even if in most cases the conditions are not as good as those of the case introduced in this paper, from the economic standpoint. Indeed these handle-filters often turn out to be absolutely necessary to reduce the pollution output while allowing the recycling of retained materials.

Basis: ton of metal oxides

Pollution balance:

	Rejects		Old Procedure	New Procedure
	Dust	kg/t		0.035
	lead oxide		70 to 90%	70 to 90%
Air	tin oxide		5 to 30%	5 to 30%
	antimony oxide		0 to 5%	0 to 5%

Economic balance: 1982 Dollars

	Old	New
Investment		119,000
Annual cost (differential)		31,000
Annual returns (differential)		160,800

48

OLD PROCEDURE

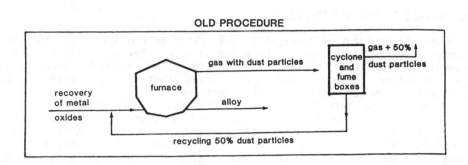

NEW PROCEDURE

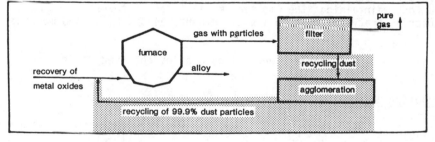

ALUMINUM FABRICATION
PROCESS CHEMISTRY MODIFICATION

ADVANTAGES OF THE NEW PROCESS

The first point of interest in the use of molten metal is for the energy savings it produces. From the point of view of environment, the results are not large. There is a reduction of noise and of combustible gas emission. However, this case shows that the improvement of the environment can have the secondary outcome of searching for energy savings or raw material savings. Consumption of energy is often accompanied by atmospheric pollution. From the economical point of view, this process is quite interesting; it is a cheaper investment and requires small operating costs. The molten metal reduces the possibilities of speculation on the metals price (storage is difficult) and makes the changing of a supplier more difficult.

POSSIBLE EXTENSIONS

The transportation of molten metal is also practiced for cast iron. The small capacities (two metric ton) bags, used for the transportation and for maintaining the molten metal, can interest the cast iron factories that use, even in a small amount, alloys by second melting.

Basis: ton of liquid alloy obtained

Pollution balance:

	Rejects		Old Procedure	New Procedure
Air	combustion gas	*TEP/t	0.081	0.014
Noise			80 decibles near the furnace	

*TEP = ton equivalent petroleum

Economic balance: 1982 Dollars

	Old	New
Investment	146,160	111,000
Annual costs	81,850	11,100
Annual returns	5,260	-

OLD PROCEDURE

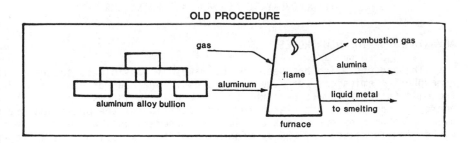

NEW PROCEDURE

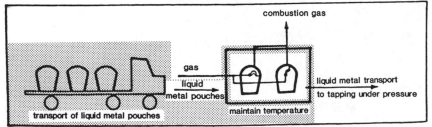

BRASS PROCESSING
OIL RECOVERY TO CONTROL AIR EMISSIONS

ADVANTAGES OF THE NEW PROCESS

The process which has been set up makes it possible to suppress the atmospheric emissions of the old process. The heat of combustion from the oils is recovered to heat the building which balances a part of the energy consumption necessary for the operation of the press and drying oven. The working conditions are improved tremendously while the metal yield is increased by 2% by reduction of the oxidation. This process also results in an improvement of the metal obtained.

POSSIBLE EXTENSIONS AND IMPROVEMENTS

Any metal shavings transformation unit can derive some benefit from the process, which often allows the rejected effluents to meet the standards but which also provides a better value of the by-products.

Basis: ton of turning in furnace

Pollution balance:

	Rejects		Old Procedure	New Procedure
Air	throughput of gas	m^3/t	4,000	17,500
		kg/t	12	0.5

Economic balance: 1982 Dollars

	Old	New
Investment	584,600	1,520,000
Annual costs	58,460	190,000
Annual returns	0	228,000

50

OLD PROCEDURE

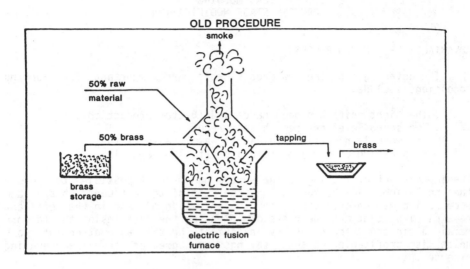

NEW PROCEDURE

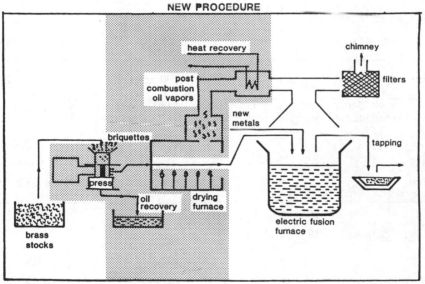

LEAD MOLDING
PROCESS STAGE MODIFICATION

ADVANTAGES OF THE NEW PROCESS

The advantages of the new process which are important for working conditions include:

. The night shift has been cancelled (better production).
. The sound level has been highly reduced.
. The work is not as painful as it was because of the suppression of carrying templates (full or empty).

The improvement of filling up the system has made it possible to reduce the pollution flow by making the treatment of the effluents cheaper and more reliable. Unfortunately, the concentration of lead in the effluent remains important and the risks of releasing the dust inside the factory remain a concern. The necessary investment for the automation of work units is profitable because savings are greater than the operating expenses.

Basis: 1000 plates

Pollution balance:

	Rejects		Old	New
Water	throughput	m^3/1000 plates	80	2
Air	smouldering emissions from lead	g/1000 plates	830	110
	smouldering workshop environment	g/1000 plates	380	16
	dust	mg/Nm^3	0.8-1	0.4-0.5
Noise		decibels	>95 db	<85 db

Economic balance: 1982 Dollars

	Old	New
Investment	29,230	438,480
Annual cost	2,424,800	2,259,600

* Cost of the renovation of old installation

OLD PROCEDURE

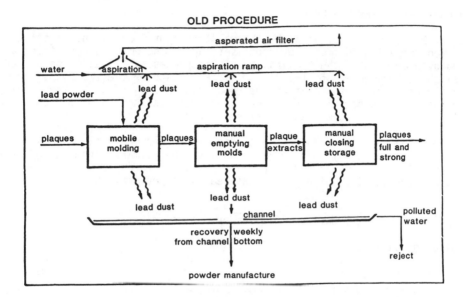

NEW PROCEDURE

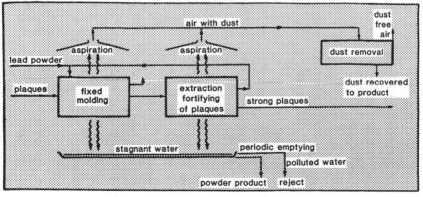

STEEL PARTS SURFACING
PROCESS MODIFICATION TO CONTROL AIR POLLUTION

ADVANTAGES OF THE NEW PROCESS

The new work room with the thermal treatment fluid baths is actually a modernization of the present set up. The working conditions are improved as much by the flexibility of the process utilization as by the healthy atmosphere of the work place. Each of the two ovens can be easily used for both heating up and cooling down with the energy consumption clearly less than before. Taking into account the characteristics of the heat carrier element (starch powder), it is possible to reach temperatures in the vicinity of $1100^{\circ}C$ ($2012^{\circ}F$), which renders the process more efficient than the former one while the working cost is reduced by one half. However, the primary advantage of this process is from the suppression of any pollution throughput while keeping a treatment quality which is at least equal to the previous one.

Basis: 1000 tempered steel rings

Pollution balance:

	Rejects		Old	New
	throughput	m^3/1000 rings	3.5	0
Water	pH	Eq/1000 rings	75 12.5	– –
Air	vapors odors dust		yes yes no	– – –
Solid Waste	Cyanide Salt (used)			0

Economic balance: 1982 Dollars
for production of 230,000 plaques per year

	Old	New
Investment	29,230	438,450
Annual cost	2,424,600	2,259,500

OLD PROCEDURE

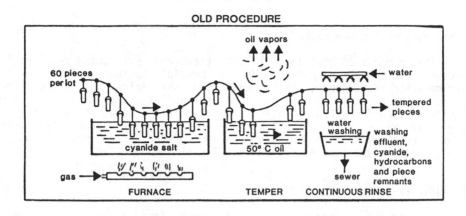

NEW PROCEDURE

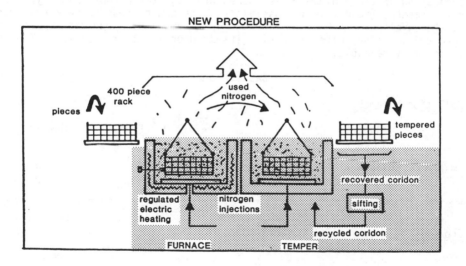

MECHANICAL WORKING
CUTTING FLUID RECYCLE

ADVANTAGES OF THE NEW PROCESS

This process of "cutting fluids" treatment put into service in Peugeot factories in Lille made it possible to substantially reduce incineration of used products and thus the associated pollution. The economic balance is positive. The reliability and longevity of the membranes used for ultrafiltration are satisfactory today (they last more than two years if they are taken care of). Starting up this treatment process does not require many modifications of the older one. Finally, energy savings are made by the elimination of incineration.

POSSIBLE EXTENSIONS AND IMPROVEMENTS

The use of "cutting fluids" in mechanical industries is very general, but small companies do not treat used products themselves. An evaluation of the minimum size necessary to start up this type of retreatment should make it possible for the new process to be widespread to the levels of mechanical industries themselves and used products recoveries. The ultrafiltration itself used in other industrial fields is now very reliable.

Basis: None

Pollution balance:

Rejects			Old Proc.	New Proc.
Water	pollution cutting liquids	m^3/yr	520	24
	COD	kg/day	250	12

Economic balance: 1982 Dollars

	Old	New
Investment	-	66,140
Annual costs	41,000	2,800

OLD PROCEDURE

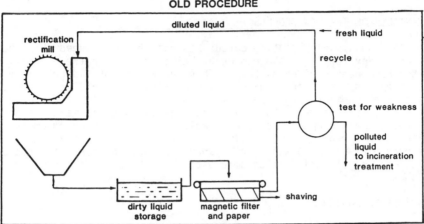

NEW PROCEDURE

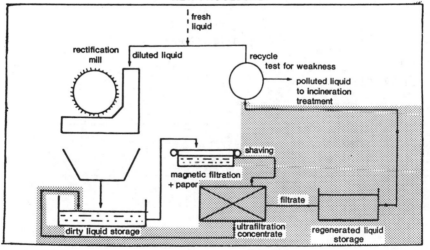

PLASTIC-METAL BONDING
REGENERATION OF PICKLING BATH

ADVANTAGES OF THE NEW PROCESS

The industrial pollution output has not been changed by the new process. Nevertheless, the number of baths treated in the detoxification center have been divided by four. Thus the amount of sludge coming from the treatment station and the risks of accidental pollution due to the transfer of chemicals have been reduced. As far as the factory is concerned, the investment is very attractive since it makes it possible to reduce the consumption of raw materials, especially chromium for which price rises rapidly, and to diminish the treatment cost of waste baths. The economics played a major role when it was decided to chose the process of regeneration of chromic baths.

POSSIBLE EXTENSIONS

This process is itself an expansion of an electrodyalizer process used for several years in order to oxidize trivalent chromium into hexavalent chromium. It can be extended to all of the units provided that the plastic materials are sufficiently clean to bond with a metal.

Basis: None

Pollution balance:

	Rejects		Old Procedure	New Procedure
Water	Running waste water bath in center of the treatment	m^3/yr.	35	9.5

Economic balance: 1982 Dollars

	Old	New
Investment	-	30,400
Annual cost	34,000	10,900

54

OLD PROCEDURE

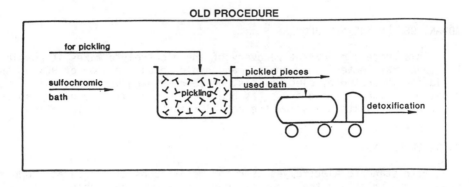

NEW PROCEDURE

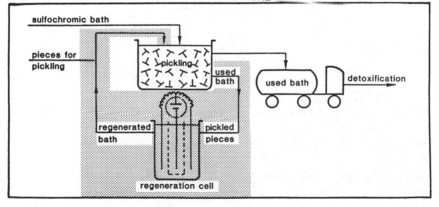

151

METAL CLEANING
ELECTROLYTIC RECOVERY OF BATHS

ADVANTAGES OF THE NEW PROCESS

This process of direct recovery of the electrolyte makes it possible to save raw materials, energy, and avoid detoxification of 90% of the pollution generated by electrolysis, while remaining very reliable. In addition to that, the process has been put into service recently, so the anticipated results are below the actual results of the installation.

POSSIBLE EXTENSIONS

With standard electrolyes used for metal deposition such as copper, brass, and nickle, the process can be very profitable. It could therefore be widespread throughout thousands of electrolysis stations in France. The techniques could be exported because it is still relatively non-existent throughout the European market.

Basis: None

Pollution balance:

	Rejects		Old Procedure	New Procedure
Water	Flow of copper and cyanides detoxiated in sodium	kg/yr	7,150	715

Economic balance: 1982 Dollars

	Old	New
Investment	** 355,000	124,000
Annual cost	*	21,000

* non-disposable but superior to the cost of the new procedure

** detoxification station for water residues from the process

OLD PROCEDURE

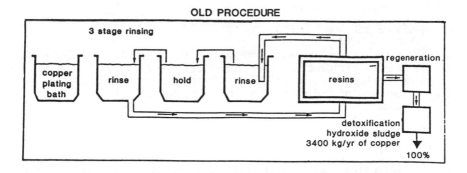

NEW PROCEDURE

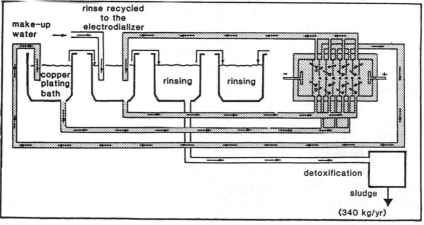

CHROME-PLATING
RINSE BATH RECYCLE

ADVANTAGES OF THE NEW PROCESS

The recycling of chromic acid makes it possible to suppress any aqueous toxic effluent and therefore sets the factory in agreement with the enforced chromic acid standards. This results in a reduction of the water needs and of the chromic acid needs, but causes extra expenses for energy and manpower.

POSSIBLE EXTENSIONS

The recycling implies that the regeneration of cationic resins and the barium sulfate sludges have to be treated. As far as the "Regie Nationale des Usines Renault" is concerned, the process is being started up on a coppering line. It can also be applied to the treatment of the liquids used to reuse the chromic passivation of parts coated with zinc or cadmium. It can also be applied after black chromium plating and to the dulling of ABS. An analogous process is used by "Trefimetaux a Oivet," to treat the reuse waters of cleaned copper parts.

Basis: 1000m2 chrome surface

Pollution balance:

	Rejects		Old Procedure	New Procedure
	throughput	$m^3/1000m^2$	60	0
Water	chromic acid	$Eq/1000m^2$ $kg/1000m^2$	250 9	0 0

Economic balance: 1982 Dollars

	Old	New
Investment	–	688.400
Annual cost	60,500	91,800

OLD PROCEDURE

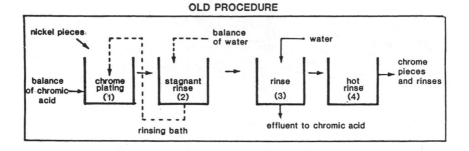

NEW PROCEDURE

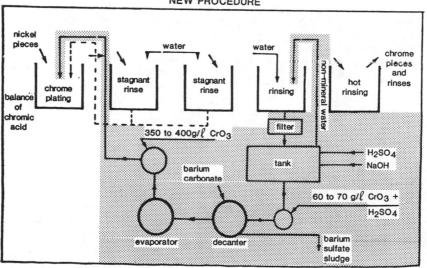

CHROME-PLATING
PROCESS MODIFICATION

ADVANTAGES OF THE NEW PROCESS

The set-up of a new chromium plating process will reduce the pollution by setting bounds to the drag out of the solution, by the small amount of water it uses, and the considerable reduction of chromium hydroxide sludges which are not soluble and which must usually be disposed of. From the economic standpoint, putting the new process into service will result in a considerable reduction of the raw materials bought. Plus, replacing the exchanger, which provides hot air by a direct utilization of the combustion gas, will make it possible to reduce energy expenses.

POSSIBLE EXTENSIONS

This treatment method can be applied to any factory having electrolytic chromium plating lines.

Basis: ton of chrome plated ion

Pollution balance:

Rejects		Old	New
cooling water (from refrigeration) m^3/t		1	0
purge Cr^{+3} in bath and steam			(purge bath
Cr^{+4}	g/t	180	(steam recycle
Cr^{+3}	g/t	14	32*
rinse water	l/t	350	recycle
Cr^{+4}	g/t	270	recycle
Cr^{+3}	g/t	20	2*
sodium sulfate	g/t	–	670
Air Cr^{+4} reject to atmosphere	g/t	2	2

* as chromium sulfate

Economic balance: 1982 Dollars

	Old	New
Investment	482,000	749,800
Annual costs	346,500	21,500

OLD PROCEDURE

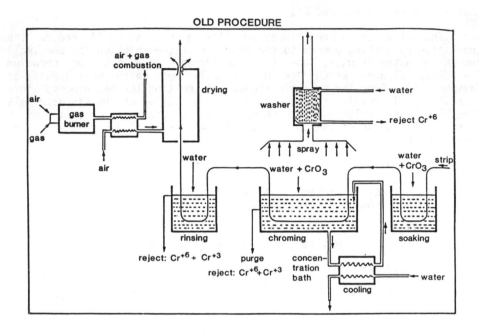

NEW PROCEDURE

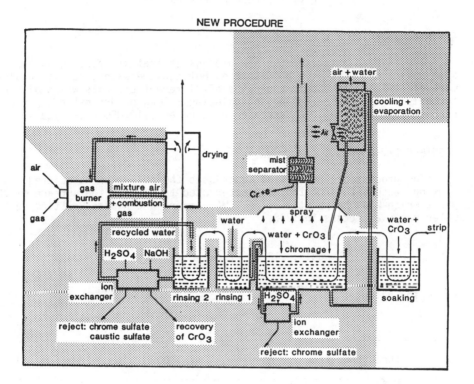

SURFACE PAINTING
RECYCLE/REUSE OF RAW MATERIAL

ADVANTAGES OF THE NEW PROCESS

The main advantage of the new process is that it suppresses nearly all pollution. The new process does not consume water, but it requires more electricity to supply the pumps of different circuits which did not formerly exist. The savings on painting (15 Kg of dry extract per day) represents a considerable decrease in expenses (110,000 F per year).

POSSIBLE EXTENSIONS

The process presently used suppresses the pollution associated with rinsing water, but not those of water used to prepare the surface of the parts which will be painted. The REMA company anticipates setting up the water recycle process in 1980.

Basis: $1000m^2$ treated surface

Pollution balance:

Rejects		Old	New
throughput from the middle m^3/1000m		32	0
MES	)	4	0
COD	) kg/1000m	32	0
Color effluent	)	obvious	-

Economic balance: 1982 Dollars

		Old	New
Investment	*	3,000	
	**	80,000	147,300
Annual cost		278,500	270,500

  * investment in 1968 Francs
 ** estimation of the investment for the installation of active carbon absorption unit

58

OLD PROCEDURE

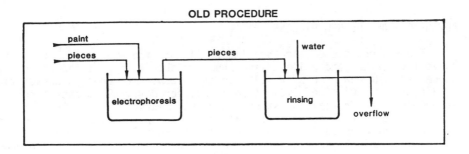

NEW PROCEDURE

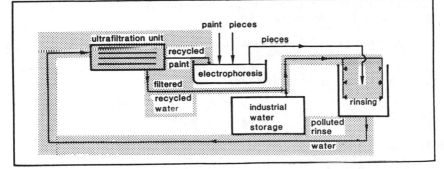

PRODUCT DRYING
VAPOR RECOVERY AND REUSE

ADVANTAGES OF THE NEW PROCESS

The new process eliminates almost all of the pollution from the solvent vapor emitted while drying the varnish. As far as the factory is concerned, the new process results in an increase of energy consumption (1.3 tons [10^3 kg] of equivalent petroleum per day). From the economical standpoint, the initial investment is rather large. The rise of oil prices makes it possible to balance the operating expenses for the recovery system by savings from the solvent recycle. This process is technically able to be profitable, but at this point it is not. In order to be profitable, the production must be increased.

POSSIBLE EXTENSIONS

Recovery of the steam over active coal is rather widespread in the French industry. But SCAL is the only one to use this sytem for acetone, methyl ethyl acetone, ethyl acetate and ethyl alcohol. Its example could encourage other industries to use the process for these products, which are generally considered hard to recover.

The process will be improved again when the problem of neutralization of alpha diketo butane in water and solvent is solved. SCAL has given up using the basic neutralization with sodium carbonate and is still studying the problem.

Basis: None

Pollution balance:

Rejects		Old Procedure	New Procedure
solvent vapors: to the maximum	kg/hr	280	14
within middle kg/t of used solvents		700	30

Economic balance: 1982 Dollars

	Old	New
Investment		1,169,700
Annual cost	1,472,800	1,446,000

162

OLD PROCEDURE

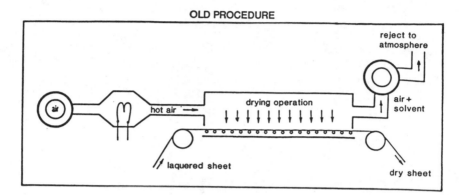

NEW PROCEDURE

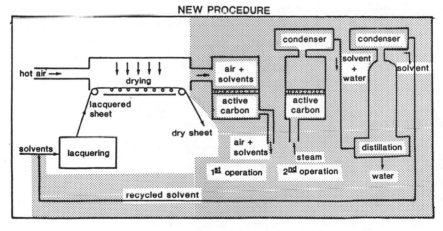

WOOL CLEANING
RECOVERY OF RESIDUES

ADVANTAGES OF THE NEW PROCESS

The new process put into service by the "Societe Dewarrin" has made it possible to totally suppress the initial pollution of the residual water and largely diminish the water consumption. This has been achieved although the cost has been half that of a standard treatment facility.

Half of the increase in energy consumption has been balanced by the combustion of the residues yielded by the effluent treatment process.

POSSIBLE EXTENSIONS AND IMPROVEMENTS

The system of treatment which has been set up is based on basic physical operations: evaporation, distillation, and combustion. When the pollution is serious and too expensive for use of the effluent treatment method, the principle of concentration destruction might be chosen in the present case.

As a matter of fact, the new process described above can be extended to all of the wool cleaning units.

Basis: ton of treated wool

Pollution balance:

	Rejects		Old Procedure	New Procedure
	throughput	m^3/t	4.5	0
Water	MES	kg/t	135	0
	BOD	kg/t	70	0
	COD	kg/t	300	0
Air	dust (particles)	kg/t	0.5	0.5
	SO_2	kg/t	2	1.5
Waste Solids	ashes	kg/t	0	100

Economic balance: 1982 Dollars

	Old	New
Investment	9,500,400	4,384,800
Annual cost	1,096,200	1,754,000
Annual returns	964,650	964,650

60

OLD PROCEDURE

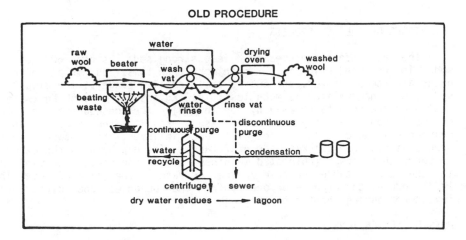

NEW PROCEDURE

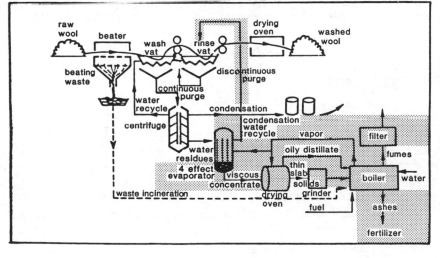

TEXTILES
RECOVERY AND RECYCLE OF RINSE BATHS

ADVANTAGES OF THE NEW PROCESS

The new process consumes much less strong soda, since 60% of the soda is recycled (in fact, much more than this; but if periodic new baths are accounted for, the 60% is the overall average). The amount of soda to be neutralized is much less: 410 Kg instead of 1450 Kg (without including the soda taken away by the thread). The consequences in terms of pollution are very obvious. As a result, the consumption of hydrochloric acid is less. The difference in investments between the two processes is very quickly made profitable. The gap between the expenses in terms of manpower (automated machines) and of the chemicals (minimized in the above table because of the increase of the price of soda since 1977) reduces the expense by one third and balances the differential investment during the first year.

POSSIBLE EXTENSIONS

The recycling of the rinse water loaded with soda can be extended to the mercerizing industries and more broadly to industries having a lot of soda discharged. Note that the mercerizing process using ammonia also makes it possible to pollute rinsing water with soda. The solution has been chosen by "les Etablissements Foucheurs" at Frelinghien (59).

Basis: ton of yarn

Pollution balance:

	Rejects		Old	New
	throughput from the middle	m^3/t	80	12
Water	MES	kg/t	0.375	0.07
	COD	kg/t	22	10
	neutralizing soda	kg/t	360	100

Economic balance: 1982 Dollars

	Old	New
Investment	330,100	435,100
Annual cost	564,100	375,100

OLD PROCEDURE

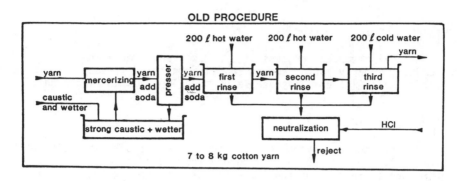

NEW PROCEDURE

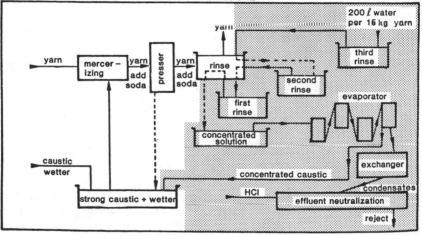

FABRIC CLEANING
OIL RECOVERY BY SOLVENT EXTRACTION

ADVANTAGES OF THE NEW PROCESS

The oil removal process in a solvent environment combines economical, energy, and ecological interests. It suppresses a large amount of manipulations and because of that improves working conditions.

POSSIBLE EXTENSIONS

Considering the advantages of this oil removal process with a solvent, it should be quickly agreed to be used by many companies that require the degreasing of fabrics except those who require a quality of flannel softness.

Basis: ton of tissue

Pollution balance:

	Rejects		Old	New
	used water	m^3/t	45	0
Water	greases	l/t	38	–
	sodium chloride)	kg/t	4	4
	)		14	–
	)		4.5	–
	formic acid	kg/t	2	–
Water	MES	kg/t	22	
	MO)		135	
	)	kEq/t	2	
	Salt	mMho/t	2600	

Economic balance: 1982 Dollars

	Old	New
Investment	429,900*	218,300
Annual cost	98,100	53,250
Annual returns		8,700

* estimations

OLD PROCEDURE

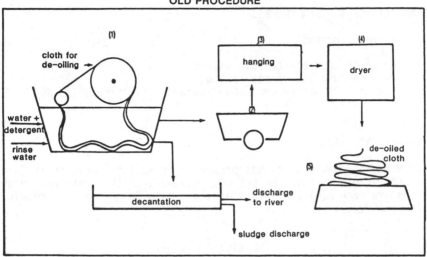

NEW PROCEDURE

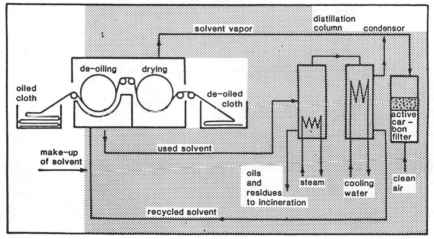

TEXTILE DYEING
RECYCLE OF INPUT DYES

ADVANTAGES OF THE NEW PROCESS

The new process:

- makes it possible to save water;
- results in a 25% energy saving;
- reduces 30% of the throughput volume. (This recycled 30% is mostly polluted.)

POSSIBLE EXTENSIONS

Since the new process has proven to be satisfactory, it is used in the fabrication center of BOLBEC. However, in this case two recovering basins are each equipped with a dyeing device in order to recover two baths per cycle.

Basis: ton treated

Pollution balance:

	Rejects		Old Procedure	New Procedure
	throughput	m^3/t	50	35
Water	MES	kg/t	non-disposable	1
	BOD	kg/t	non-disposable	10
	COD	kg/t	non-disposable	35
				0.05

Economic balance: 1982 Dollars

	Old	New
Investment	–	180,250
Annual cost	36,000	2,970

OLD PROCEDURE

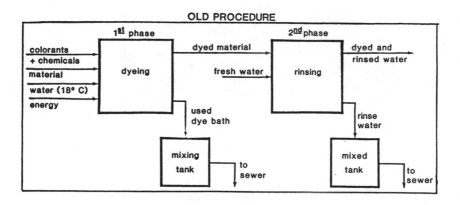

NEW PROCEDURE

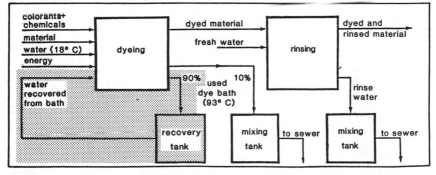

TEXTILE DYEING
WATER RECYCLE AND EFFLUENT REDUCTION

ADVANTAGES OF THE NEW PROCESS

The process setup is the "Descamps Demeestiere" factory has the main aspects of:

o A modification of the setup which makes it possible to
   achieve a better dyeing yield and therefore less waste.

o A separation of the waste streams depending on their level
   of pollution and on the temperature. Each corresponds
   to a specific orientation of the new technologies.

o Reduction of the amount of effluent.

o Better treatment of these effluents (and energy savings).

For the entire unit, the result achieved is important since the amount of polluted water is divided by three. The pollutants are also reduced.

The economic balance is positive. The variable expenses have been reduced by 22% making the investments made over three years profitable. These results will be improved by setting up the exchanger which will reduce the energy consumption by 15%.

POSSIBLE EXTENSIONS

The process of drying by vacuum exhausting is now efficient only in the case of rolled up and wrapped textiles.

In spite of this restriction, the technological improvements will certainly broaden the use and extension of this process to many drying factories. The modifications on the autoclaves are obviously so specific to that kind of activity and cannot be transferred directly to others. Although, the effluent separation, depending on their pollution level, is a principle which can be generalized to many activities.

Basis: ton of treated cotton

Pollution balance:

	Rejects		Old Procedure	New Procedure
	throughput	m^3/t	>400	150
Water	COD	kg/t	>350	160
	BOD	kg/t	>85	40
	MES	kg/t	>50	6

Economic balance: 1982 Dollars

		Old	New
Investment		non-disposable	1,870,800
Annual cost		2,383,300	1,832,900

OLD PROCEDURE

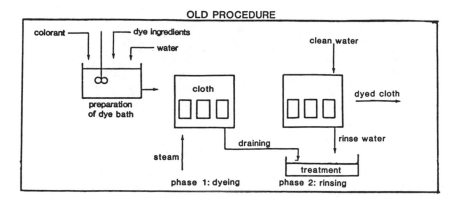

NEW PROCEDURE

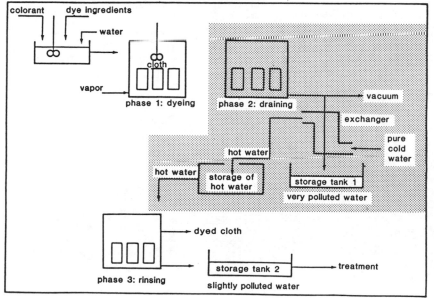

TEXTILE DYEING
PROCESS CONTROL IMPROVEMENTS

ADVANTAGES OF THE NEW PROCESS

From the pollution standpoint, the advantage of the new process is obvious:
- the amount of "White Spirit" discharged in the water is divided by eight.
- the amount of "White Spirit" discharged in the air is divided by 3.5.
- the amount of coloring paste thrown out is divided by two.

The economic interest is also very obvious in spite of a small investment (*) (most of the investment concerns the paste preparation plant). In return, the dyeing of the impression paste is longer with the new process (the amount of water it contains is larger) which results in a higher energy consumption.

POSSIBLE EXTENSIONS

This process can be applied to any field of production using coloring on fabrics, papers, etc. It is just beginning to be widely used.

* The difference in investments is pointed out in the economic balance.

$$Basis: \ 1000m^2 \ of \ tissue$$

Pollution balance:

	Rejects		Old	New
Water	white liquor	$1/10^3m^2$	57	7.5
Air	white liquor vapors	$kg/10^3m^2$	175	50
	paste color	$kg/10^3m^2$	57	28

<u>Economic balance</u>:   1982 Dollars

		Old	New
Investments	*	961,400	998,850
Annual costs		807,400	713,650

** Actual investment corresponds to the renovation of the old process.

OLD PROCEDURE

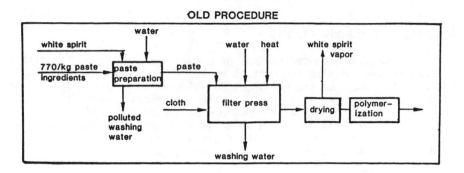

NEW PROCEDURE

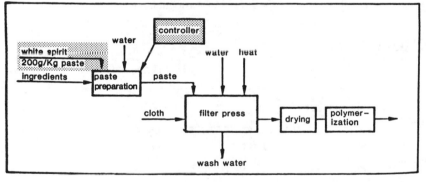

TANNERY
CHROME RECOVERY AND REUSE

ADVANTAGES OF THE NEW PROCESS

Because of the concentrations of organic materials, it is difficult
to reduce the chromium ion content on the resin. On the other hand, the
standard treatment of the effluents do not make it possible to eliminate
these chromium ions. In return, the new process makes it possible to
reduce the amount of chromium contained by the tanning effluent by 60%.
Since it also provides recycling in the manufacturing, lowering the
supplying cost of chromium for the factory. This process guarantees a
good purification and is financially promising.

POSSIBLE EXTENSIONS

This process can be adaptable to every chromium tanning factory.

Basis: ton tanned hide

Pollution balance:

	Rejects		Old Procedure	New Procedure
	throughput	m^3/t	60	60
Water	MES	kg/t	135	100
	BOD	kg/t	77	75
	COD	kg/t	210	200
	Chrome ions	kg/t	7.5	3
	Sludge	m^3/t	0.6	0.4
			29% chg mat'l	27% chg mat'l

Economic balance: 1982 Dollars

	Old	New
Investment	-	69,500
Annual costs (differential)	200,400	53,000

66

OLD PROCEDURE

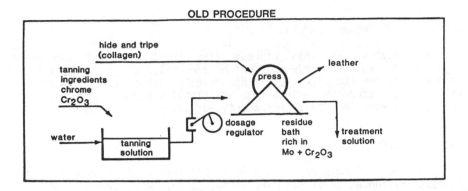

NEW PROCEDURE

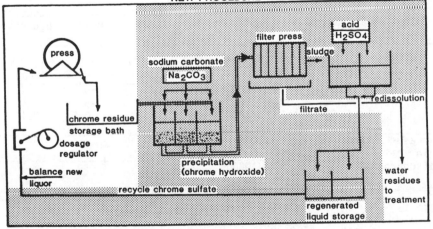

TANNERY
RECOVERY OF GREASE AND OIL

ADVANTAGES OF THE NEW PROCESS

The interest of this process lies in the elimination of all
significant solvent rejects, which are recycled in manufacturing.
Economically the recycle makes it possible to highly reduce the amount of
solvent purchased but brings some additional expenses in energy, which,
in fact, do more than balance the previous savings.

POSSIBLE EXTENSIONS

The process could be widespread to all treatment of solvents in an
aqueous environment. It has already been accepted by several tanneries.

Basis: 1000 hides

Pollution balance:

	Rejects		Old Mill	New Procedure Mill	Degreaser
	throughput	$m^3/1000$	249	247	1.75
Water	MES	kg/1000 hides	175	56	0.4
	MO	kg/1000 hides	135	135	11
	toxins	kEq/1000 hides	5	0.5	0.1
greases in	throughput	1/1000 hides	-	-	15
discharge	BOD	kg/1000	-	-	2.1
	MES	kg/1000	-	-	0.05

Economic balance: 1982 Dollars

	Old	New
Investment	-	48,600
Annual cost (differential)	-	2,400

OLD PROCEDURE

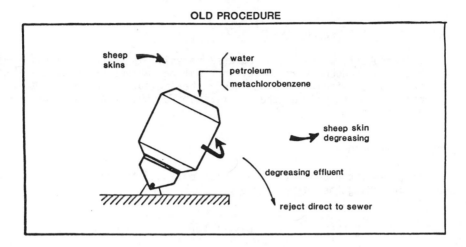

NEW PROCEDURE

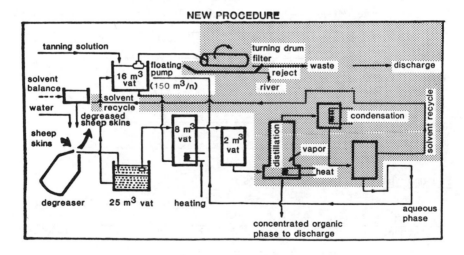

PRESSED BOARD MANUFACTURING
RECYCLE OF EFFLUENTS

ADVANTAGES OF THE NEW PROCESS

This process, which has integrated recycle, is so far the only means
by which to reduce the pollution that has a reasonable cost. Plus,
contrary to purification processes, the pollution almost totally
disappears and therefore the utilization of this process is desirable
from any standpoint.

POSSIBLE EXTENSIONS

This new technique has already been patented in many countries which
use it, including the United States.

Basis: ton

Pollution balance:

	Rejects		Old Procedure	New Procedure
	throughput	l/t	30,000	5 to 130
Water	MES	kg/t	40	0.6
	BOD	kg/t	35	0.1
	COD	kg/t	130	0.4
	MO	kg/t	70	0.2
	MO & MES	kg/t	110	0.8

Economic balance: 1982 Dollars

		Old	New
Investment	*	1,461,500	1,000,000
Annual cost		558,100	73,100

* estimation of the cost of the effluent treatment station

68

OLD PROCEDURE

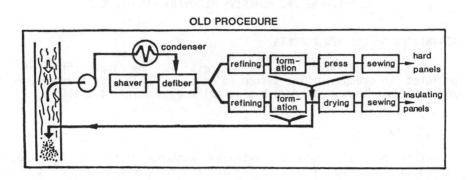

NEW PROCEDURE

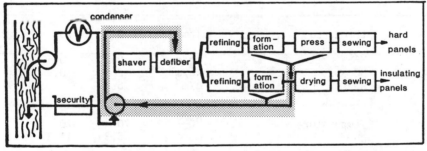

PLYWOOD FABRICATION
EFFLUENT AND ADHESIVE RECOVERY AND RECYCLE

ADVANTAGES OF THE NEW PROCESS

The first advantage of the process is the suppression of any polluted throughput. The system is quite reliable. In any case, it is more reliable than a standard biological treatment plant station.

POSSIBLE EXTENSION

This process could be extended without difficulty to other manufacturers that produce plywood. The recycle process of the resins might be extended to other manufacturing using phenolic glues.

Basis: m3 of laminated paper

Pollution balance:

	Rejects		Old		New
	throughput	$1/m^3$	300	20	0
Water	MES	kg/m^3	0.1	0.5	-
	MO	kg/m^3	2	0.8	-
	PO_4	g/m^3	6	-	-
	Temperature	^{o}C	70	20	-
	pH		5.3	11.3	-
	phenolic resin		-	3.2% dry wt.	-
	calcium carbonate/organic chg.		-	2.5% in wt.	-

Economic balance: 1982 Dollars

	Old	New
Investment	409,250	587,800
Annual costs	460,400	235,900

69

OLD PROCEDURE

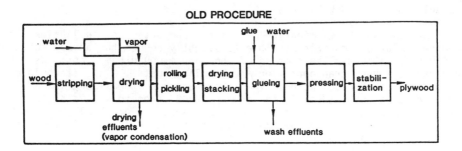

NEW PROCEDURE

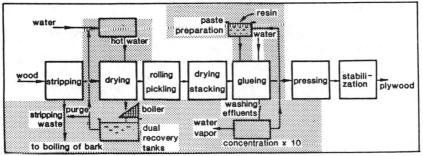

PAPER MILL
ENHANCED FIBER UTILIZATION

ADVANTAGES OF THE NEW PROCESS

Because of the elimination of a large proportion of pollution, the chosen solutions make it possible to slightly reduce the annual costs of effluent treatment. The storage of the sludge would have cost as much because of the transportation and handling costs. Plus, it would only have displaced the pollution. The incineration of the sludge in a lime oven also makes it possible to recover about three tons (10^3 Kg) per day of calcium carbonate. Using this product instead of Kaolin makes it possible to consume a material which is produced locally instead of an imported material.

The chosen processes have been set by the necessity of reducing pollution. They also permit savings for the "Aussedat-Rey" company.

POSSIBLE EXTENSIONS

Such a process, which substitutes carbonate for kaolin and incinerates the primary sludges in a lime oven, can be set up in any integrated Kraft-factory.

At "Saillat-sur-Vienne", it has been necessary to modify the paper manufacturing process (for instance, the adhesion which was previously performed in an acid medium must now be performed in a neutral medium).

These obstacles can be regarded as minor ones compared with the general advantages of the process.

Basis: ton of writing paper

Pollution balance:

	Rejects		Old Paste	Old Paper	New Paste	New Paper
	throughput	m^3/t	250	25	250	15
Water	MES	kg/t	6	15	6	3.5
	BOD	kg/t	5	2	5	0.5
	COD	kg/t	95	4	95	3
Solids	Sludge		180 22% dry material		0	

Economic balance: 1982 Dollars

	Old	New
Investment	3,215,500*	2,952,400**
Annual cost	935,400	613,870
Annual returns	–	–

* Furnace for incineration of sludge with kaolin and treatment of
   paper effluents
** Furnace for lime modification and paper effluent recycle

OLD PROCEDURE

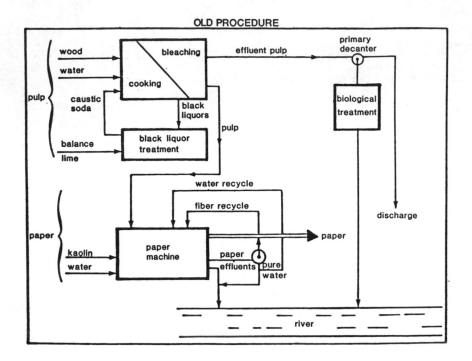

NEW PROCEDURE

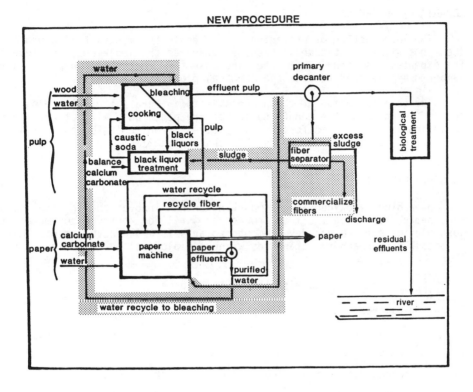

PULP AND PAPER INDUSTRY
OXYGEN BLEACHING

ADVANTAGES OF THE NEW PROCESS

The introduction of whitening by oxygen in the manufacturing process does not imply noticeably extra expenses at the implementation level. The expenses resulting from the increase of energy consumption are balanced by the profits from the chemical profits.

The main advantage of the process is to integrate the bleaching of the effluents to the manufacturing of the paper pulp without any other pollution as is the case for the old process of whitening by contact with chlorine and soda followed by the bleaching with lime.

POSSIBLE EXTENSIONS

Bleaching by oxygen is an important French innovation in the pulp and paper technology. The process can still be improved and makes it possible for some future developments to arise. These developments make it possible to reach the standards without biological purification.

Basis: ton paste to paper

Pollution balance:

	Rejects		Old	New
	throughput of bleaching effluent	m^3/t	100	50
Water	MES	kg/t	10	*
	BOD	kg/t	25	*
	COD	kg/t	75	*
	color	$kg(Pt/Co)/t$	90	75

* little modification

Economic balance: 1982 Dollars

	Old	New
Investment	3,663,000	11,819,300
Annual cost	995,500	734,500

OLD PROCEDURE

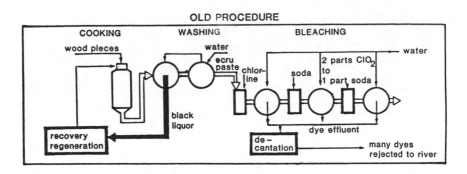

NEW PROCEDURE

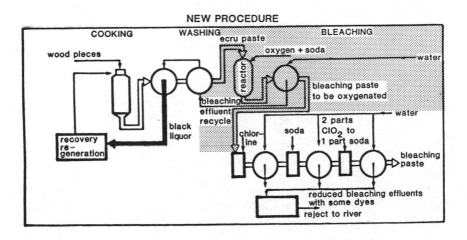

195

PAPER MANUFACTURING
COMPLETE PULPING EFFLUENT RECYCLE

ADVANTAGES OF THE NEW PROCESS

The transition from the closed loop paper machine creates corrosion problems in the pipes and results in rather high maintenance expenses. However, this change in the process has made it possible to reduce the water consumption, recover the fibrous and raw mineral materials, save thermal energy, and suppress any pollution throughput.

POSSIBLE EXTENSIONS

The process consists of total or partial closing of the water circuits which can be applied with a few modification to all of the paper factories which manufacture paper of a similar quality.

Basis: ton of paper product

Pollution balance:

	Rejects		Old Procedure	New Procedure
	throughput	m^3/t	56	0
Water	MES	kg/t	31	0
	COD	kg/t	19	0
	BOD	kg/t	9	0
	MO	kg/t	12	0

Economic balance: 1982 Dollars

	Old	New
Investment	803,400	428,500
Annual costs	non-disposable	15

OLD PROCEDURE

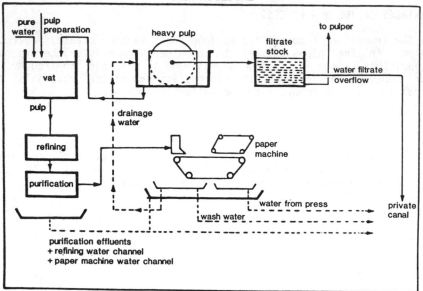

72

NEW PROCEDURE

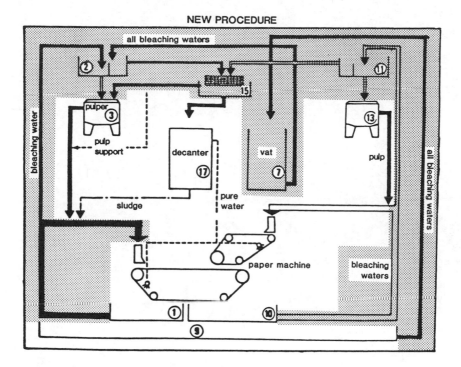

199

PULP AND PAPER PLANT
EFFLUENT RECYCLE AND FIBER RECOVERY

ADVANTAGES OF THE NEW PROCESS

With the separation of waste streams, the different effluents have various levels of pollution. This makes it possible to accurately:

o select them for an immediate recovery;

o take the necessary measures to remove the polluting element.

POSSIBLE EXTENSIONS

In this process, the recovering circuit can be applied (in every case) at a reasonably important level. As for the sewer charges, the possible use of sludge depends on the location of a user because the cost of transportation remains high compared to the low price of sludge (145 Francs per ton).

Basis: ton of paper

Pollution balance:

		Rejects	Old Procedure	New Procedure
	throughput	m^3/t	60	40
Water	MES	kg/t	11.5	1.8
	BOD	kg/t	4.5	1.5
	COD	kg/t	11	3

Economic balance: 1982 Dollars

	Old	New
Investment (differential)		233,850
Annual costs (differential)	18,700	23,970
Annual returns (differential)		12,600

73

OLD PROCEDURE

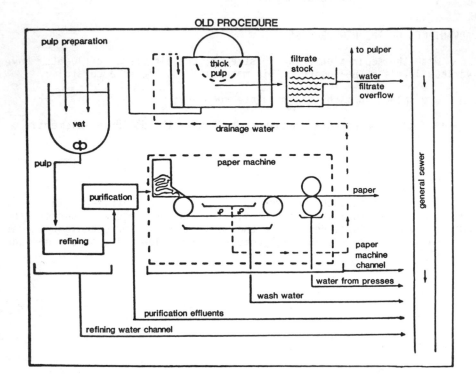

202

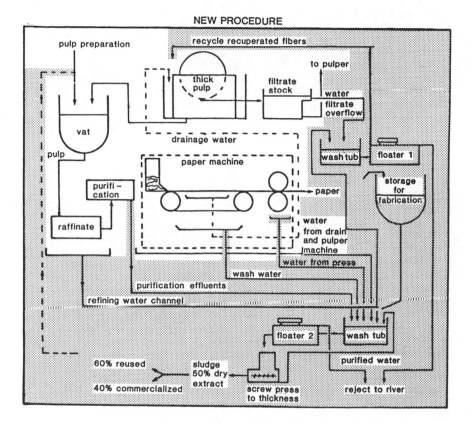